Capture and Reuse of Project Knowledge in Construction

H.C. Tan

C.J. Anumba

P.M. Carrillo

D. Bouchlaghem

J. Kamara

C. Udeaja

WILEY-BLACKWELL

A John Wiley & Sons, Ltd., Publication

This edition first published 2010
© Hai Chen Tan, Chimay Anumba, Patricia Carrillo, Dino Bouchlaghem,
John kamara, Chika Udeaja

Blackwell Publishing was acquired by John Wiley & Sons in February 2007.
Blackwell's publishing programme has been merged with Wiley's global Scientific,
Technical, and Medical business to form Wiley-Blackwell.

Registered office
John Wiley & Sons Ltd, The Atrium, Southern Gate, Chichester, West Sussex, PO19
8SQ, United Kingdom

Editorial office
9600 Garsington Road, Oxford, OX4 2DQ, United Kingdom
2121 State Avenue, Ames, Iowa 50014-8300, USA

For details of our global editorial offices, for customer services and for information
about how to apply for permission to reuse the copyright material in this book
please see our website at www.wiley.com/wiley-blackwell.

The right of the author to be identified as the author of this work has been asserted
in accordance with the Copyright, Designs and Patents Act 1988.

Library of Congress Cataloging-in-Publication Data
Capture and reuse of project knowledge in construction/ Hai Chen Tan ... [et al.].
        p. cm.
    Includes bibliographical references and indes.
    ISBN 978-1-4051-9889-9 (hardback : alk. Paper) 1. Construction industry –
        Information technology. 2. Knowledge management. 3. Intellectual capital.
        I. Tan, Hai Chen.

    TH215.C37 2010
    690.068—dc22

                                                                    2009023945

A catalogue record for this book is available from the British Library.

Set in 9.5/12.5pt Palatino by MPS Limited, A Macmillan Company, Chennai, India
Printed in Malaysia

1 2010

Contents

Author Details

Chimay Anumba is Professor and Head of the Department of Architectural Engineering at the Pennsylvania State University, USA. His research interests include construction engineering, advanced engineering informatics, concurrent engineering and knowledge management – fields in which he has over 400 publications. Professor Anumba's work has received support worth over £15m from a variety of sources. He has also supervised more than 33 doctoral candidates and 16 postdoctoral researchers. In recognition of his substantial and sustained original contributions to the field of Construction Engineering and Informatics, Professor Anumba was awarded the higher doctorate degree of Doctor of Science, DSc, by Loughborough University in July 2006. In January 2007, he was also awarded an Honorary Doctorate (Dr.h.c.) by Delft University of Technology in The Netherlands for outstanding scientific contributions to Building and Construction Engineering.

Dino Bouchlaghem is a Professor of Architectural Engineering at Loughborough University and Director of the Engineering Doctorate Centre for Innovative and Collaborative Construction Engineering. He has research interests and expertise in sustainable design and construction, knowledge management and Design for Safety and Security. He led numerous projects funded by the Engineering and Physical Sciences Research Council, the Department of the Environment Transport and the Region, the British Council and the EU. He set up and coordinated an Architectural Engineering Task Group for the International Council for Building Research Studies. He is the editor-in-chief of the *International Journal of Architectural Engineering and Design Management*.

Patricia M. Carrillo is Professor of Strategic Management in Construction in the Department of Civil and Building Engineering at Loughborough University. Her areas of expertise are knowledge management, business performance measurement and management, IT in Construction and disaster resilience in the built environment. She was awarded the prestigious Royal Academy of Engineering Global Award. She was also visiting professor at the University of Calgary, Canada and University of Colorado, USA. To date she has published books and over 160 journal papers, conference papers and reports.

John Kamara is a Senior Lecturer in the School of Architecture, Planning and Landscape at Newcastle University, UK, and was formerly coordinator of the CIB Task Group on Collaborative Engineering. He is a registered facilitator in the use of the Design Quality Indicator (DQI) tool in evaluating the design and construction of buildings. His research and engagement activities, which have received support from the UK government and private industry, are in the areas of project development, knowledge management, collaborative working in construction and construction informatics. He has over 90 publications in these fields.

Hai Chen Tan is an Assistant Professor in the Department of Built Environment at Universiti Tunku Abdul Rahman (UTAR), Malaysia. Prior to joining UTAR, he worked as a research associate for an EPSRC-funded knowledge management research project at Loughborough University where he obtained his MSc in Construction Project Management and PhD. His research interests include improving the performance of construction companies through knowledge management and the use of ICT, and other current issues in construction. He is a member of the Chartered Institute of Building (CIOB).

Chika Udeaja graduated as a Civil Engineer and worked briefly as a site engineer and as a design engineer before undertaking postgraduate studies in Concrete structures at Imperial College London and this was followed by a brief spell as a bridge engineer in Malaysia. He returned to the UK, to undertake a postgraduate research in Construction management at the London South Bank University. On completion of his PhD in 2003, he joined University of Newcastle as a researcher, and was involved in developing CAPRIKON and other research projects. He is currently a senior lecturer in the School of the Built Environment at Northumbria University, UK.

Foreword

Considerable knowledge is generated during the course of a construction project. Sadly, only a small fraction of this is captured and an even smaller fraction is subsequently reused. This problem is often associated with the fragmented nature of the construction industry, with each project involving a variety of disciplines and organisations. It has also been a major contributor to the inefficiencies associated with the construction industry. It is, therefore, imperative in seeking to improve the productivity of the industry and the profitability of the firms that operate within it, that this new knowledge is captured, shared and reused. Attempts are now being made by many companies to capture the learning on projects through post-project reviews and various 'lessons learned' automated data files but these have only been marginally successful. A major limitation of these approaches is that the review sessions take place long after the learning event, and many of the details and subtleties are not captured in the automated data files, making it difficult for participants to fully recall and utilize the details of the lessons learned or the context in which they were learned.

This book provides guidance on how the learning on projects can be captured during the course of a project (i.e. live), with a view to reusing the new knowledge at the later stages of the same project or in new projects. It provides guidance on how to ensure that the lessons learned are shared between the members of the project team, and across corporate enterprises without a significant administrative burden. The key elements of the approach developed are encapsulated in a software tool that will prove invaluable to design and construction organisations. Some of the excellent features of the tool, which are not adequately provided for in most knowledge management systems, include: the primary focus on reusable knowledge (thereby ensuring that only the most useful knowledge is captured), effective support for knowledge management at the project organisation level and the provision of the 'context for reuse' for each knowledge entry.

I strongly recommend this book to anyone working in the construction industry. The fundamental principles contained in the book are based on extensive research and will also be useful to professionals in other project-based industries.

William M. Brennan

Executive Vice President

Turner Construction Company

November 2009

Acknowledgements

We are grateful to the Engineering Physical Sciences Research Council (EPSRC) for funding the research on which this book is based. We acknowledge the contributions of the various organisations that collaborated on the project. We are also indebted to our families, whose continued love and support enabled us to put this book together.

1 Introduction

This chapter briefly introduces the background and justification for the 'live' capture and reuse of project knowledge. It outlines the importance of the methodology developed and provides a summary of the book's objectives and a guide to its contents.

1.1 Background

According to Drucker (1993), we have entered the age of the knowledge economy where knowledge has sidelined both capital and labour to become the 'sole factor of production'. In a knowledge economy, knowledge is regarded as the single most important asset of organisations (Stewart, 1997). An organisation's competitive advantage lies in the knowledge residing in the heads of its employees and the capability to harness the knowledge for meeting its business objectives, for continuous improvement and for avoiding the repetition of past mistakes (Davenport et al., 1997; Demarest, 1997; Drucker, 1998; Bollinger and Smith, 2001). Related to this, some companies have started to audit the value of their knowledge and include this information in the annual report to stakeholders (Davenport et al., 1997).

Given the growing importance of knowledge towards the success and even the survival of an organisation, it is not surprising that the significance of a systematic or organised knowledge management (KM) approach is being increasingly recognised. KPMG's (2003) survey results revealed that the KM practice in the organisations surveyed had improved from one mainly characterised by the lack of an established implementation strategy in 1998 to one approaching a higher maturity level with greater board/management support in 2002/2003. In the context of the construction industry, a survey of construction organisations revealed that about 80% perceived KM as having the potential to provide benefits to their organisations (Carrillo et al., 2003). However, in terms of implementation, KM in the industry is still at its infancy with various shortcomings in the practice for managing knowledge relating to and arising from a project (Khalfan et al., 2002). The rationale for this book

hence stems from the need to provide guidance to organisations interested in the management of knowledge within a project environment,
such as in the construction industry. It focuses on the 'live' capture and
reuse of project knowledge. The reasons for this are discussed below.

1.2   The need for live capture and reuse of project knowledge

The shortcomings of KM practices in construction are closely related to
the industry's characteristic of a predominantly project-based industry. A
typical construction project involves many people and organisations with
different specialisations or expertise forming a virtual organisation for
the duration of a project. These projects are usually unique, very complex
and require the combined knowledge and expertise of all the project team
members in order to deliver the project successfully. Hence, it is not surprising to find that most of the knowledge of the construction industry is
generated in projects, by staff belonging to different disciplines, during
the process to deliver a custom-built facility in accordance with the client's requirements and business objectives.

The knowledge generated from a project can be about the best practices learned on how to carry out tasks in a more efficient way, or some
negative lessons learned which have led to losses and slowed down the
progress of the project. The ability to manage the knowledge generated
from the projects (including the capture of project knowledge and its
subsequent transfer) not only can help to prevent the 'reinvention of the
wheel' and the repetition of similar mistakes, but also serves as the basis
for innovation and overall improvement. This is crucial in view of the
fact that knowledge, particularly the lessons learned, is actually acquired
from both the positive lessons learned and the mistakes made at a cost to
the organisation. However, recent evidence has revealed that the ability
to learn from within and across projects is critical but difficult to achieve
(Kamara *et al.*, 2003). This is mainly due to the following reasons:

- In a project, each individual only knows bits of the whole story of the
  project (Kerth, 2000). Knowledge created in a project is scattered in
  the memory of various project team members but none retains a complete set of the knowledge created. Therefore, when the virtual organisation or the project team formed for a project is disbanded upon the
  completion of the project, the knowledge retained by each member is
  likely to be minimal. Most of the knowledge gained from the project
  is not shared and is therefore lost.
- Some companies have tried to address the aforementioned knowledge loss problem by conducting post project reviews (PPRs) after the
  completion of a project so as to capture the knowledge gained or the
  lessons learned. However, the success of PPRs is often undermined

by the lack of time for conducting it as other project team members may be transferred to and therefore involved in new projects. The effectiveness of PPRs in facilitating the capture and reuse of knowledge learned is also affected by the lack of a suitable format for representing the knowledge captured, and a mechanism for sharing the knowledge captured across projects for reuse. In addition, humans are not without weaknesses and this is particularly so when it comes to memorising facts (Ebbinghaus, 1885). The time lapse in capturing the knowledge gained through PPRs and the current practice of condensing the knowledge into bullet points have led to the loss of important details about the knowledge (Kamara et al., 2003).

- The reassignment of individuals or even the whole project team from one project to another as an attempt to transfer the knowledge acquired makes organisations vulnerable when there is a high staff turnover (Kamara et al., 2003). This is substantiated by the persisting high staff turnover rate, which was 20.2% in 2003, in the UK's construction industry (CIPD, 2004). In addition, this method does not proactively facilitate the sharing of knowledge acquired from a project with others who are not involved in that project. Furthermore, it also suffers from the aforementioned human weakness in memorising facts.
- Reluctance to share knowledge amongst the project team members due to commercial sensitivity, corporate restrictions as to the sharing of information and knowledge (Barson et al., 2000) and the fact that the organisations collaborating in one project may actually compete elsewhere (Kamara et al., 2003).

One potential solution for the above problems could be a methodology that is capable of:

- Facilitating and encouraging project team members to share important knowledge;
- Storing the knowledge learned in a format that helps the sharing and understanding of its content;
- More importantly, enabling the capture and reuse of knowledge in real time (i.e. 'live') or as soon as possible after the knowledge is created to address the knowledge loss problem due to time lapse in capturing that knowledge.

The importance of 'live' capture of knowledge is supported by the recent survey of construction and client organisations involved in PFI (Private Finance Initiative) projects where it was identified as crucial by over 70% of the organisations (Robinson et al., 2004). Kamara et al. (2003) contend that a methodology that facilitates the 'live' capture and reuse of project knowledge allows the knowledge captured from the initial stages of a project to be reused at subsequent stages of a project (intra-project knowledge transfer), and helps to ensure that a more complete set of project knowledge is captured. Using the term 'information' synonymously with 'knowledge', McGee (2004) also states that the capture and presentation of real-time 'information' is crucial in helping to:

- Prevent mishaps from happening owing to the capability to share lessons learned and critical information in real time;
- Seize the opportunities to reuse the knowledge captured by making knowledge available for reuse once it is captured;
- Maximise the value of reusing knowledge, particularly if the benefit brought about through reusing the knowledge is time-related.

A review of the existing literature indicates that a number of research projects have been undertaken to help improve the management of knowledge in construction and other project-based industries. These research projects focused only on either specific types of knowledge [e.g. C-SanD (2001)], specific project phases [e.g. KLICON (McCarthy *et al.*, 2000)], specific types of construction organisations [e.g. SMEs in Boyd *et al.* (2004)] or strategic issues of managing knowledge in construction [e.g. CLEVER (Kamara *et al.*, 2003)]. The need for an approach which is capable of the 'live' capture of project knowledge, however, has not been adequately addressed. This book therefore addresses the importance of developing a methodology that facilitates the 'live' capture and reuse of project knowledge in construction and other project-based industries.

1.3   The objectives and contents of the book

This book covers the development of a methodology for the 'live' capture of reusable project knowledge that reflects both the organisational and human dimensions of knowledge capture and reuse, as well as exploiting the benefits of technology. The 'live' capture of reusable project knowledge is defined in this context as the capture of knowledge as soon as possible after it is created or identified. This methodology was developed in response to the various shortcomings of current practices in managing project knowledge (previously outlined) and the benefits offered by the ability to facilitate the 'live' capture, sharing and reuse of project knowledge within a dynamic and challenging project environment. The background study, development, testing and evaluation of the methodology are described in the various chapters of this book as follows:

- *Chapter 1 – Introduction*: This chapter provides the background to the studies that led to the writing of this book. It justifies the need for a methodology for the 'live' capture and reuse of project knowledge in construction and other project-based industry sectors, and introduces the contents of the book.

- *Chapter 2 – Knowledge Management – Key Concepts*: This chapter reviews the definition of knowledge, the different perspectives and processes of KM, shortcomings of current practice for knowledge capture and reuse in construction, KM research projects in construction and the importance of the 'live' capture and reuse of project knowledge in construction.
- *Chapter 3 – Knowledge Capture and Reuse*: This chapter presents the reviews of the potential types of reusable project knowledge in construction, the learning situations where most of the new learning are created, the current practice for the capture of knowledge focusing on the capability to facilitate the 'live' capture of project knowledge and the soft (organisational, cultural and human) issues that affect knowledge capture and reuse.
- *Chapter 4 – Collaborative Learning*: This chapter reviews the concept of Collaborative Learning (CL) and discusses its importance in a project environment. Drawing on the construction industry as an example of a project-based industry, it explores how CL can be implemented in project teams and presents the benefits of this approach.
- *Chapter 5 – Knowledge Reuse Requirements*: This chapter presents case studies on the current approaches for knowledge capture, and the end-users' requirements for knowledge capture and reuse. The development of the methodology that facilitates the 'live' capture and reuse of project knowledge based on the case study findings is also explained.
- *Chapter 6 – Development and Operation of a 'Live' Capture Methodology*: This chapter presents the structure of the 'live' capture and reuse of project knowledge framework, and the system architecture, software development as well as the operation of the prototype application. The results of the evaluation are presented and analysed in detail.
- *Chapter 7 – Conclusions and Recommendations*: This chapter brings together the findings and draws conclusions from the book. It also discusses further research that can be conducted to enhance the methodology and the functions of the prototype software application.

2 Knowledge Management – Key Concepts

This chapter reviews some key knowledge management (KM) concepts including: definition of knowledge, the different perspectives and processes of KM, shortcomings of current practice for knowledge capture and reuse in construction, KM research projects in construction and the importance of the 'live' capture and reuse of project knowledge in construction.

2.1 Defining knowledge

In the context of KM, knowledge is defined in various ways reflecting different research perspectives. Most of the definitions of knowledge fall into two categories: knowledge can be defined by comparing or relating it to data and information (Marshall, 1997; Burton-Jones, 1999; Kanter, 1999), or knowledge can be defined as knowledge per se (i.e. without directly linking knowledge to information and data) (Nonaka and Takeuchi, 1995; OECD, 1996; Rennie, 1999; Davenport and Prusak, 2000).

In the former case, knowledge is considered as an entity which is at a higher level and authority than data and information (Stewart, 1997). Data is said to be a set of discrete facts about events (Davenport and Prusak, 2000), while information is 'data endowed with relevance and purpose' (Drucker, 1998) which can be created by adding value to data through contextualising, categorising, calculating, correcting and condensing it (Davenport and Prusak, 2000). Knowledge can then be described as 'actionable information' (O'Dell et al., 1998; Tiwana, 2002) which 'gives one the power to act, to make decisions that are value producing' (Kanter, 1999). In the real world, however, a clear-cut distinction between knowledge, information and data is not always possible as the differences between these terms are just a matter of degree (Davenport and Prusak, 2000). Furthermore, depending on the relevance of the knowledge and knowledge base (KB) of individuals, knowledge for one person may be interpreted as information to others and vice versa (Bhatt, 2001).

The second perspective defines knowledge as knowledge per se (i.e. by depicting knowledge's characteristics, quality and constituents rather than contrasting it with information and data). Hence, it avoids the intriguing distinction between knowledge and information in particular. An important example within this category is Davenport and Prusak's (2000) definition of knowledge as 'a fluid mix of framed experience, values, contextual information, and expert insight that provides a framework for evaluating and incorporating new experiences and information'. Apart from this, knowledge is also defined as a series of know-what, know-how and know-who (OECD, 1996; Rennie, 1999), a 'dynamic human process of justifying personal belief towards the truth' (Nonaka and Takeuchi, 1995) and the product of learning (Orange *et al.*, 2000). The definition by Davenport and Prusak (2000) which has captured the various subtle features of knowledge is hence preferred.

2.2   Knowledge management

KM generally deals with the systematic and organised attempt to use knowledge within an organisation to transform its ability to store and use knowledge to improve performance (KPMG, 1998). There is a plethora of definitions for KM available, all attempting to encapsulate what KM is and how it should be done (Quintas *et al.*, 1997; O'Leary, 2001; Diakoulakis *et al.*, 2004; Nicolas, 2004), but no consensus has, hitherto, been reached. The perspectives of KM which are most relevant to the contents of this research are as follows:

- Functionalist vs. interpretivist (Venters, 2002);
- Information systems vs. human resource management;
- Interdisciplinary perspective (Jashapara, 2004);
- Soft and hard approaches (Kamara et al., 2003).

Other perspectives include radical humanism and radical structuralism perspectives (Schultze, 1998), process-centred and product-centred perspectives (Mentzas *et al.*, 2001), contingency perspective (Becerra-Fernandex and Sabherwal, 2001), process and interaction views (Rollett, 2003), artefact-oriented, process-oriented and autopoietic-oriented epistemology perspectives (Christensen and Bang, 2003). The aforementioned four most relevant perspectives are described in the following section.

2.2.1   *Functionalist vs. interpretivist*

Applying Burrel and Morgan's (1979) framework of social and organisational inquiry, Schultze (1998) identified four paradigms of KM research, namely radical humanism, radical structuralism, interpretivism and functionalism, as shown in Figure 2.1.

Within the paradigms, there is a continuum between the subjective and objective perspectives. From the objective perspective, knowledge is

The sociology of radical change

S u b j e c t i v i t y	Radical humanism	Radical structuralism	O b j e c t i v i t y
	• Knowledge as the social practice of knowing. • Value of knowledge and work is contested and serves as a source of conflict.	• Knowledge as an object that can exist independently of human action and perception. • Value of knowledge and work is contested and serves as a source of conflict.	
	Interpretivism	Functionalism	
	• Knowledge as the social practice of knowing. • Consensus about the value of knowledge and work.	• Knowledge as an object that can exist independently of human action and perception. • Consensus about the value of knowledge and work.	

The sociology of regulation

Figure 2.1   Four paradigms in KM research (*Source*: Schultze, 1998)

considered as an object waiting to be discovered and which can exist in a variety of forms (e.g. tacit and explicit) and in a variety of locations (e.g. in the individual, group or organisation) (Schultze, 1998). From the subjective perspective, it is asserted that knowledge is continuously shaping and being shaped by the social practices of communities, and cannot be located in any one place because it cannot exist independent of human experience and social practices of knowing (Schultze, 1998). In addition, these paradigms can also be contrasted by analysing how 'knowledge work' and the value of associated knowledge are viewed. From the sociology of regulation perspective, knowledge work is deemed necessary following the shift towards the knowledge economy and the value of knowledge is acknowledged (Schultze, 1998). However, from the sociology of radical change perspective, it is asserted that knowledge work is just 'another development in the political economy of capitalism' and knowledge is devalued through 'technologization' (Schultze, 1998).

Among the four paradigms, current research in KM is dominated by functionalism which is frequently contrasted with interpretivism as there is a 'paucity of radical structuralist or humanist perspectives in knowledge management research' (Jashapara, 2004). The weight of both the radical structuralist and humanist perspectives is very likely being affected by their inability to accommodate the post-structural theories (Schultze, 1998). The aforementioned paradigms are therefore being combined into a 'critical perspective' to accommodate the post-structural theories (Schultze, 1998). Venters (2002) disregards the radical structuralist and humanist perspectives, and concentrates only on functionalist and interpretivist perspectives:

- *Functionalist perspective*: Apart from inheriting the characteristics of the objective perspective, the functionalist approach is highly scientific, employing accounting methods, codification and structures to exploit knowledge, and depends heavily on technology and 'database-led activities' to achieve these objectives (Venters, 2002).

- *Interpretivist perspective*: This approach inherits the characteristics of subjective perspective and focuses on supporting the social structures and processes within which knowledge is shared (Venters, 2002). This perspective does not view technology as the solution by itself, but rather as support to the social activity of sharing knowledge (Venters, 2002).

2.2.2 Information systems vs. human resource management

The current definitions of KM are predominantly from the information systems and human resource management perspectives (Jashapara, 2004), which correspond to the technocratic and behavioural schools of KM proposed by Earl (2001). From the information systems perspective, KM is concerned with the use of information and communication technology (ICT) to facilitate the capture, deployment, access and reuse of information and knowledge (O'Leary, 2001), whereas the human resource management perspective emphasises the establishment of means to motivate and facilitate knowledge workers to develop, enhance and use their knowledge in order to achieve organisational goals (Beijerse, 1999). However, leveraging knowledge through ICT alone is often hard to achieve (Siemieniuch and Sinclair, 1999; Walsham, 2001; Rollett, 2003) as there are human, cultural and organisational issues such as reluctance to share knowledge which are not readily resolved by ICT. Conversely, a purely human resource management approach is not going to benefit from the faster, cheaper and broader source of data and means of communication to enable people to generate and share knowledge offered by ICT. Therefore, it is argued that an integrated approach of KM combining information systems (technology) and human resource management (people) synergised by the benefits of both perspectives is likely to be a more viable option (Davenport, 1998).

2.2.3 Interdisciplinary perspective

Jashapara (2004) contends that KM has its roots in various disciplines, namely anthropology, economics, sociology, strategy, management science, human resource management, information science, philosophy, psychology and computer science. It is therefore argued that an integrated, interdisciplinary and strategic perspective of KM is necessary for a KM initiative to succeed (Jashapara, 2004). Based on this assertion, Jashapara (2004) groups the various KM disciplines into four dimensions (see Figure 2.2), that is:

- strategy
- organisational learning
- systems and technology
- culture

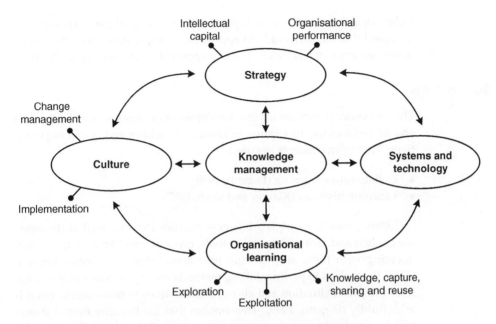

Figure 2.2   Dimensions of KM (*Source*: Adapted from Jashapara, 2004)

Jashapara (2004) argues that the strategic purpose of KM is to increase intellectual property and enhance organisational performance. Organisational learning, which comprises individual, group and organisational level learning (Crossan *et al.*, 1999), is the process of improving actions through better knowledge and understanding (Fiol and Lyles, 1985) within an organisation. In order to fully explore and exploit knowledge, systems and technology are crucial to the facilitation and enhancement of the cycle of knowledge creation, capture, organisation, evaluation, storage and sharing (Jashapara, 2004). In addition to the systems, and technology and organisational processes, the interdisciplinary perspective also addresses the crucial cultural and change management dimensions for the implementation of KM as many well-planned initiatives have been futile because of overlooking these dimensions (Jashapara, 2004).

2.2.4   Combined soft and hard approaches

A combined 'soft' (i.e. organisational, cultural and people issues) and 'hard' (ICTs) approach is introduced by Kamara *et al.* (2003) for the 'live' capture of knowledge in construction. The main feature of the 'live' capture methodology is the capability to facilitate the capture of knowledge once it has been created or identified. This combined soft and hard approach adopts a pragmatic view acknowledging that there are strengths and shortcomings in the KM practice solely focused on either technological (i.e. hard) or organisational and cultural (i.e. soft) issues. It is argued that the soft and hard approaches complement each other and a combined approach is therefore more appropriate (Kamara *et al.*, 2003).

Soft concepts

The soft concept focuses on the development of organisational processes and procedures for the capture of knowledge within and across organisations. Two main concepts are used:

- Collaborative learning (Digenti, 1999);
- Learning histories (Kleiner and Roth, 1997).

Collaborative learning is a business practice that is aimed at discovering explicit and tacit collaboration tools, processes and knowledge, experimenting with them and creating new knowledge from them (Digenti, 1999). It employs experimentation, methods and approaches that emerge from the preset situation and allows organisations to move across boundaries fluidly (Digenti, 1999). This ensures that the learning from a group, which can also be a construction project team, is transferred back to the organisation (Kamara *et al.*, 2003). This is discussed in greater detail in Chapter 4.

A learning history is a process for capturing usable knowledge from an extended experience of a team and transferring that knowledge to another team that may be distant in terms of context (Dixon, 2000). Kamara *et al.* (2003) argue that although construction projects and the teams that implement them are unique, the team structure, processes and skills involved in these projects are similar, and these provide the opportunity for the reuse of knowledge. Using the concept of learning history, the learning of one team (from critical events on a project) can therefore serve as a catalyst to a similar team to deal with issues in a different context (Kamara *et al.*, 2003).

Hard concepts

The hard concepts include the available ICT applications that are currently being used in the construction industry, particularly project extranets, workflow management tools and other groupware applications for collaborative working (Kamara *et al.*, 2003).

Project extranets are dedicated Web-hosted collaboration and information spaces for the architectural, engineering and construction industry that support design and construction teams (Augenbroe *et al.*, 2002; Kamara *et al.*, 2003). Utilising Web-based technology, project extranets allow distributed team members to collaborate, as well as to share, view and comment on project-relevant information without the need to meet in one location (Kamara *et al.*, 2003). Kamara *et al.* (2003) argue that the growing use and the collaborative facilities provided by project extranets make them a suitable platform on which a methodology for 'live' capture of knowledge can be mounted. Due to the current limitations of project extranets (e.g. being purely document-centric with limited facilities for workflow), the proposed methodology will be complemented by other ICT such as workflow modelling and automation tools (Kamara *et al.*, 2003).

If the soft and hard approaches are analysed individually, they are closely related to the information systems (or technocratic) and human resource management (or behavioural) perspectives respectively. However, the combined soft and hard approach resembles the integrated perspective proposed by Jashapara (2004) and presents a more balanced approach than that offered by the solution from either of the two extremes. This perspective appears to incline towards the functionalist perspective as it considers knowledge as something that can exist independently of humans and that can be captured using ICT.

The issues pertaining to the collaborative learning and learning histories are explored in detail in Chapter 4. The following section presents the benefits and barriers to KM.

2.3 Benefits and barriers to KM

The growing importance of KM is often related to the emergence of the knowledge-based economy and the importance of knowledge in providing competitive advantage (Drucker, 1993; Beijerse, 1999; Bollinger and Smith, 2001). KPMG (1998) indicates that KM can lead to:

- Better decision-making;
- Faster response time to key issues;
- Increased profit;
- Improved productivity;
- Creation of new/additional business opportunities;
- Reduced costs;
- Better sharing of best practice;
- Increased market share and share price;
- Better staff attraction and retention.

Other benefits identified include:

- Reduction of rework, and continuous improvement and better sharing of tacit knowledge (Carrillo *et al.*, 2004);
- Improved efficiency in project implementation (CBPP, 2004);
- More effective discovery and access of knowledge (Egbu and Botterill, 2001).

Although the benefits of KM are apparent, its implementation may not be so straightforward and trouble-free. Very often, it is undermined by main barriers that prevent the full leverage of the benefits. According to Carrillo *et al.* (2004), the barriers to KM implementation are:

- Lack of standard work processes;
- Not enough time;
- Organisational culture;
- Not enough money;
- Employee resistance;
- Poor IT infrastructure.

Corresponding to the findings by Carrillo *et al.* (2004), KPMG (1998) also identifies the lack of time, standard work processes or skills in KM, funding, appropriate technology and a supporting culture as the main barriers to KM implementation. Other than this, KPMG's (1998) findings reveal that about a quarter of the respondents mention the lack of commitment from the senior management as a barrier to the implementation of KM. The lack of senior management commitment is critical as the implementation of KM is time consuming and may entail huge investment of organisational resources. Furthermore, the attempt to address the aforementioned barriers (such as the modification of existing organisational culture to one that is supportive of KM activities and the provision of sufficient funding for new IT infrastructure) for the implementation of KM is less likely to be successful without continuous commitment from the senior management.

2.4 KM processes

Different researchers have used different terms for the same KM processes or stages (see Appendix A). What differentiates each of the models is the difference in perspectives, focus and level of detail. Bhatt (2001) delineates the sequence of the KM processes as: knowledge creation, knowledge validation, knowledge presentation, knowledge distribution and knowledge application. However, there is evidence that KM processes may not exist in that linear sequence (Demarest, 1997). Demarest (1997) notes that there can be iterations between the KM processes, such as that between the embodiment (i.e. presentation) and dissemination (i.e. distribution) of knowledge. His study also reveals that some of these stages may exist simultaneously, such as in the case of construction (the process of discovering and structuring knowledge) and use of knowledge, where people may have put the knowledge into practice while it is being 'constructed'. The KM process models also differ in the levels of detail: some of which do not take into consideration the issue of knowledge obsolescence in KM (Demarest, 1997; Soliman and Spooner, 2000;

Table 2.1   Relationship between the capture, reuse and maintenance of knowledge

Live Capture and Reuse of Project Knowledge (CAPRIKON)	Robinson et al. (2001)	Kululanga and McCaffer (2001)	Bhatt (2001)	Rollett (2003)
Capture • Identifying • Locating • Representing • Storing • Validating	• Discovering • Locating • Capturing • Organising • Storing	• Acquiring • Creating • Storing	• Creating • Presentation • Validating	• Planning • Creating • Assessing • Integrating • Organising
Sharing • Sharing	• Sharing • Transferring	• Sharing	• Distributing	• Transferring
Reuse • Adapting • Applying	• Modifying • Applying	• Utilising	• Applying	
Maintain • Archiving • Retirement	• Archiving • Retirement			• Maintaining

Kululanga and McCaffer, 2001) and some do not address the need to validate the knowledge.

Four main KM processes (see Table 2.1), which have incorporated the notions of knowledge obsolescence and validation, are proposed based on the KM process models that are developed within the context of construction (Kululanga and McCaffer, 2001; Robinson *et al.*, 2001):

- Knowledge capture;
- Knowledge sharing;
- Knowledge reuse;
- Knowledge maintenance.

2.4.1 Knowledge capture

Knowledge capture comprises three sub-processes:

- *Identifying and locating knowledge*: This deals with the identification of the types/categories of knowledge to be managed, and the location of learning situations (Kamara *et al.*, 2003) where most of the new knowledge is created and people with the knowledge required. Knowledge can be captured internally within an organisation (e.g. conducting an internal review) or externally (e.g. by recruiting staff from other companies) (Kululanga and McCaffer, 2001), and through 'creating new knowledge' or collating 'already existing knowledge' (Rollett, 2003).
- *Representing and storing knowledge*: This encompasses indexing, organising and structuring knowledge (Goodman and Chinowsky, 2000; Robinson *et al.*, 2002; Rollett, 2003) into theme-specific knowledge areas (Maier, 2002), and authoring knowledge (Markus, 2001) in the

standard or format specified with the details required, adding context to the knowledge depicting where the knowledge was generated and used, where the knowledge may be useful and the conditions for reuse (Hansen and Davenport, 1999).

- *Validating knowledge*: Knowledge validation often refers to the verification and evaluation processes of the knowledge base but there is evidence that it is also a crucial process in KM (Bhatt, 2001; Mach and Owoc, 2001). In the context of KM, it is argued that validation is likely to focus on (albeit not restricted to) explicit or codified knowledge instead of the tacit knowledge which is notoriously difficult to articulate and capture. Validation of knowledge may comprise the following:
  - ○ *Verification*: Like information, the accuracy, correctness and completeness of knowledge captured need to be verified before it is shared or transferred for reuse.
  - ○ *Evaluation*: The pertaining question is whether the knowledge entered is important and reusable. Only important and reusable knowledge should be captured in order to prevent and reduce the knowledge overload problem (Kamara *et al.*, 2003).

Validation of knowledge is intended to ensure the credence of knowledge captured, and that the knowledge captured is stored with all the relevant contextual details and in the format required.

2.4.2 Knowledge sharing

This is about the provision of the right knowledge to the right person at the right time (Mertins *et al.*, 2001; Robinson *et al.*, 2002) or within the shortest time possible. This process can be passive, such as publishing a newsletter or populating a knowledge repository for users to browse, or active, such as 'pushing' knowledge via an electronic alert to those who need to know (Markus, 2001), which may also be known as knowledge-pull and knowledge-push (Rollett, 2003: p. 83) respectively. Dixon (2000) has recognised five types of knowledge transfer (i.e. serial, near, far, strategic and expert transfer), based on who the intended receiver is, the nature of the tasks and the types of knowledge to be transferred. The details are summarised as follows (Dixon, 2000):

- *Serial transfer* is a process that moves the unique knowledge that each individual has constructed into a group or public space so that the knowledge can be integrated and made sense of by the whole team.
- *Near transfer* is the replication of knowledge learned by a team to other teams that are doing very similar work.
- *Far transfer* is very similar to near transfer, except that far transfer is non-routine and the knowledge concerned or to be transferred is mainly tacit.

- *Strategic transfer* is the transfer of the crucial collective knowledge (both tacit and explicit) of an organisation in order to accomplish a strategic task that occurs infrequently but is of critical importance.
- *Expert transfer* is applicable when teams facing an unusual technical problem beyond the scope of their own knowledge seek the expertise of others in the organisation to help them address it.

Knowledge transfer can also happen between people (e.g. meetings and conferences), person to computer (e.g. knowledge bases and expert systems) and computer to computer (e.g. data mining and intelligent agents) (Skyrme, 1998). Although the tools and methods used are dominated by ICT applications (Mertins *et al.*, 2001), effective knowledge sharing is also underpinned by a supportive organisational culture and trust between the people involved (Newell *et al.*, 2002).

2.4.3 Knowledge reuse – adapting and applying

This covers the reuse of knowledge through the re-application of knowledge, such as the re-application of best practice as mentioned by Szulanski (2000), and the reuse of knowledge for innovation with necessary adaptation or integration (Egbu *et al.*, 2001; Majchrzak *et al.*, 2004). The reuse of knowledge through adaptation involves re-conceptualising the problem and searching for reusable ideas (i.e. knowledge), scanning and evaluating reusable ideas, analysing the ideas in depth and selecting the best idea and developing fully the reused idea, which may ultimately lead to innovation (Majchrzak *et al.*, 2004).

2.4.4 Knowledge maintenance – archiving and retirement

Knowledge may become obsolete over time (Pakes and Schankerman, 1979; Rich and Duchessi, 2001). The development of a discipline often constitutes new information, rules and theories, which may render part of the old information, rules, theories and hence the relevant knowledge obsolete (Nonaka and Takeuchi, 1995; Bhatt, 2001). In addition, when new sets of tools and technologies, and processes and procedures are employed by an organisation, these also often result in the need to update and refine the skills of its employees so that they can swiftly switch to the new competitive realities (Bhatt, 2001). This process covers reviewing, correcting, updating and refining knowledge to keep it up to date, preserving and removing obsolete knowledge from the archive (Rollett, 2003).

2.5 KM in construction

There is evidence that the importance of KM has been recognised by the construction industry. A survey of UK project-based organisations shows that about 50% of the respondents (majority were from the construction industry) noted that KM would result in new technologies and new processes that will benefit the organisations (Egbu, 2002). This finding is supported by another survey of construction organisations which reveals that about 40% already had a KM strategy and another 41% planned to have a strategy within a year (Carrillo *et al.*, 2003). Furthermore, about 80% also perceived KM as having the potential to provide benefits to their organisations, and some had already appointed a senior person or group of people to implement their KM strategy (Carrillo *et al.*, 2003). Despite the awareness of the importance of KM to the industry revealed in the above studies, there are still limitations identified in current practice for the capture and reuse of project knowledge in the industry (Kamara *et al.*, 2003). These are discussed below.

2.5.1 *Shortcomings of current practice*

It has been identified that the overall processes of KM in the construction industry (architecture, engineering and construction) are characterised by the following:

- Most of the construction knowledge resides in the minds of individuals working within their specific domain (Khalfan *et al.*, 2002).
- The knowledge gained is often poorly organised and there are seldom processes in place for disseminating useful knowledge to other projects (Khalfan *et al.*, 2002).
- The intent behind decisions is often not recorded or documented. There is difficulty in tracking the people involved in a decision-making process and who understand the context of making the decision for the purpose of knowledge sharing (Khalfan *et al.*, 2002).
- There is a strong reliance on the knowledge accumulated by individuals, but no formal way of capturing and reusing much of this knowledge (Kamara *et al.*, 2002b).
- The use of long-standing (framework) agreements with suppliers to maintain continuity (and the reuse and transfer of knowledge) in the delivery of projects for a specific client (Kamara *et al.*, 2002b).
- The capture of lessons learnt and best practice, such as in the operational procedures and design guidelines, which serve as a repository of process and technical knowledge. Post project reviews (PPRs) are usually the means for capturing lessons learned from projects (Kamara *et al.*, 2002b).
- The involvement (transfer) of people in different activities as the primary means by which knowledge is transferred and/or acquired (Kamara *et al.*, 2002b).
- The use of formal and informal feedback between providers and users of knowledge as a means to transfer learning/best practice, as well as

to validate knowledge (e.g. site visits by office-based staff to obtain feedback on work progress) (Kamara *et al.*, 2002b).

- A strong reliance on informal networks and collaboration, and 'know-who' to locate the repository of knowledge (Kamara *et al.*, 2002b).
- Within firms with hierarchical organisational structures, there was a reliance on departmental/divisional heads to disseminate knowledge shared at their level, to people within their sections (Kamara *et al.*, 2002b).
- The use of appropriate IT tools (such as GroupWare, Intranets){AQ1} to support information sharing and communication (Kamara *et al.*, 2002b).

Kamara *et al.* (2002b) note that the heavy reliance on knowledge accumulated by individuals, PPRs and specific contractual/organisational arrangements (e.g. framework agreements) are considered the key approaches for direct transfer of project knowledge. However, the shortcomings impeding the effective capture and reuse of project knowledge are observed in the aforementioned approaches (Kamara *et al.*, 2003).

Post project reviews

This is the most common approach used in the industry for the capture of learning (Orange *et al.*, 1999). The shortcomings of PPRs identified by Kamara *et al.* (2003) are:

- Insufficient time is often allocated for the review to be conducted effectively (if conducted at all), as relevant personnel would have moved on to other projects.
- It does not allow the current project to be improved by incorporating the lessons being learnt as the project progresses.
- Loss of important information or insights due to time lapse in capturing the learning.
- In consolidating the learning of people involved, it is not an effective mechanism for the transfer of knowledge to non-project participants.
- The learning captured is limited in scope as the perspective is that of members within only one of the participating organisations in the project (i.e. it is not collaborative).

These shortcomings of PPR are explored in detail in Chapter 3.

Reliance on people for the transfer of knowledge

Kamara *et al.* (2003) note that the reliance on people, based on the assumption that the knowledge acquired from one project can be transferred to another project by that individual when s/he is reassigned to another project, makes organisations vulnerable when there is a high staff turnover. This is critical in view of the persisting high staff turnover rate, which was 20.2% in 2003, in the UK construction industry (CIPD, 2004). Furthermore, the transfer and sharing of knowledge through this method is very likely to be limited to the people who are working together with the knowledge provider in the project. Other projects, other members of staff within the organisation but not involved in the project and those located at other offices may therefore not benefit from this method. The availability of the knowledge provider and the relationship between the knowledge provider and the knowledge receiver are also likely to influence the willingness of the knowledge provider in sharing his/her knowledge.

In addition, humans are not without weaknesses and this is particularly so when it comes to memorising facts (Ebbinghaus, 1885). The problem of the loss of important information or insights due to the time lapse in capturing the learning through post-project evaluation is in fact due to the weakness of human memory. As the transfer of knowledge through reassignment of people is also heavily dependant on human memory, it is not surprising that it still suffers from the same knowledge loss problem as in the case of PPRs.

Contractual and organisational arrangements

The dominant culture of competitiveness and the fact that construction organisations collaborating in one project may actually compete in another project have made the construction organisations reluctant to share critical knowledge or to divulge secrets to others, as that might weaken their competitive advantage (Kamara *et al.*, 2003). Therefore, even though the use of long-standing framework agreements (e.g. within a partnering contract) with suppliers to maintain continuity in the delivery of projects for a specific client is designed to ensure that the learning is reused in future projects, there is still no guarantee that the learning of individual firms is shared with other participants in the agreement (Kamara *et al.*, 2003).

Commercial sensitivity and security of knowledge is another critical issue and barrier to inter-organisational knowledge capture and reuse which involves a number of organisations with different business objectives (Barson *et al.*, 2000). Corporate security restrictions imposed on posting of information/knowledge have further added to the problem (Ardichvili *et al.*, 2003) as people have been indirectly discouraged from sharing their knowledge especially where the boundary of such restrictions is not made clear.

2.5.2   *KM research projects in construction*

In view of the numerous shortcomings of KM current practice in construction and hence the ample room for improvement, it is not surprising that a number of research projects have been undertaken in this area. In the United Kingdom, some of these include:

Cross-sectoral Learning in the Virtual Enterprise (CLEVER)

This project aimed to derive generic structures for KM practices and to develop a framework for the transfer of knowledge in a multi-project environment in construction (Kamara et al., 2002a). The framework developed assists construction firms in articulating their KM problems and in selecting an appropriate strategy for the transfer of knowledge that is appropriate to their organisational and cultural contexts (Kamara et al., 2003).

Knowledge Management for Improved Business Performance (KnowBiz)

The aim of the project was to establish the relationship between KM practices and business performance in construction firms (Carrillo and Anumba, 2000). A KM framework which enables organisations to link their KM initiatives to improved business performance (IMPaKT) was developed (Carrillo et al., 2003), and has been encapsulated in a software tool.

Creating, Sustaining and Disseminating Knowledge for Sustainable Construction: Tools, Methods and Architecture (C-SanD)

This project was focused on the development of a mechanism, which includes a software tool, to facilitate the capture, retrieval and creation of knowledge pertaining to sustainability in construction (C-SanD, 2001). Key outputs from the project included the development of a 'Sustainability Management Activity Zone (SMAZ)' for the Process Protocol, the development of a Web-based portal for sustainable construction knowledge and the use of soft systems methodology to identify critical issues in the management of sustainable construction knowledge.

Building a Higher Value Construction Environment (B-Hive)

This project aimed to develop processes and systems to enhance organisational learning between construction project partners (B-Hive, 2001). B-Hive developed a Cross-Organisational Learning Approach (COLA), which comprises innovative processes for review, evaluation, feedback and organisational learning supported by an information system (B-Hive, 2001).

Knowledge and Learning in CONstruction (KLICON)

This project investigated the role of IT in capturing and managing knowledge for organisational learning on construction projects (KLICON, 2001). The research also explored how detailed IDEF0 models of construction activities and information models in EXPRESS can enhance understanding of generic construction knowledge and specific project knowledge

(McCarthy *et al.*, 2000). The focus was on the passing on of knowledge about the project from early design to detailed design stages and to the contractor (McCarthy *et al.*, 2000).

Methodology, tools and architectures for electronic COnsistent *knowledGe maNagement across prOjects and between enterpriSes in the construction domain (e-COGNOS)*

This EU-funded project was aimed at specifying and developing an open model-based infrastructure and a set of tools that promote consistent KM within collaborative construction environments (Whetherill *et al.*, 2002).

An approach to KM for SMEs

This project aimed to improve KM in small- and medium-sized enterprises (SMEs) in the construction industry. The pilot study of the research involved recording aspects of the managers' personal knowledge and thinking about problem-solving events on a weekly basis using a dictaphone. The managers were then debriefed about the set of their recorded events every month in order to explicate the embedded knowledge and transform it into knowledge accessible to a wider audience (Boyd *et al.*, 2004).

A knowledge transfer approach to continuous improvement on PFI projects

This project was aimed at identifying the scope for improvement and knowledge transfer on Private Finance Initiative (PFI) projects (Robinson *et al.*, 2004). It identified the critical problem areas on PFI projects, and explored the KM issues that contribute to those problems. It also formulated a set of guidelines for enhanced knowledge transfer on PFI projects.

Benchmarking KM practice in construction

The project's primary objectives were to provide a deeper understanding of successful KM programmes and the approaches used to successfully overcome the challenges, and to identify effective ways to improve both the short- and long-term competitiveness of participating companies, all through benchmarking the activities of the group's members (Dent and Montague, 2004). A report which sets out the methodology used and the findings of the study under three areas (i.e. strategy, processes and tools) and measurement and application was published.

Business case for KM: Guidance & toolkit for construction

This project aimed to provide good practice guidance and a supporting management toolkit for practitioners to develop business plans and metrics for KM within their company (CIRIA, 2004). The outputs of the project were tailored to the needs of the target audience and the specific business context and promote opportunities for performance improvement by adopting KM practices in the UK construction industry, by raising aware-ness of the benefits of KM and by enhancing confidence of construction organisations to apply such practices (CIRIA, 2004).

Sharing knowledge between aerospace and construction

This project aimed to investigate the extent to which managerial practices can be shared between the aerospace and construction sectors (Green *et al.*, 2004). In addition, it also sought to develop an approach to knowl-edge sharing that could be implemented as part of a KM initiative within individual companies (Green *et al.*, 2004).

The literature reveals that the aforementioned research projects are focused at either:

- Strategic and business perspectives (CLEVER, KnowBiz, 'Business case for knowledge management: guidance & toolkit for construc-tion' and 'Benchmarking Knowledge Management Practice in Construction').
- Specific types of knowledge, that is knowledge pertaining to sustain-ability (C-SanD, 2001), PFI projects (Robinson *et al.*, 2004) and man-agement practice (Green *et al.*, 2004) and sustainable competitiveness (www.knowledgemanagementuk.net).
- Specific project phases, that is KLICON which focused on the transfer of knowledge from early design to detailed design stages and to the contractor (McCarthy *et al.*, 2000).

Or

- Specific type of construction organisation, for example SMEs in Boyd *et al.* (2004).

The need for an approach which is capable of capturing project knowl-edge, irrespective of the type of project, the type of construction organisa-tion and project phases and particularly capturing the knowledge 'live' (Kamara *et al.*, 2003), has not been adequately addressed. Research at Stanford University (Reiner and Fruchter, 2000) is considered as being closest to the goal of 'live' capture and reuse of project knowledge. However, the research does not cover the entire project but focuses only on the design evolution stage. The importance of a 'live' methodology proposed by Kamara *et al.* (2003) to address the limitations of current practice is discussed in detail in the next section.

2.6 The Importance of 'live' capture and reuse of project knowledge

Kamara *et al.* (2003) contend that in order to overcome the limitations in current industry practice on the capture and reuse of knowledge, it is necessary that learning from projects is captured while it is being executed (i.e. 'live') and presented in a format that will facilitate its reuse both during and after the project, and in other contexts such as professional education and training of new construction staff. The imperative of 'live' capture of knowledge is supported by the recent survey result of construction and client organisations involved in PFI projects where the 'live' capture of knowledge is noted as crucial by 76% and 70% of construction and client organisations respectively (Robinson *et al.*, 2004). Hari *et al.* (2005) noted that the speed of technological advancement requires construction organisations to 'quickly' capture, assimilate and use their knowledge in order to remain competitive. Furthermore, the need for 'live' capture of knowledge has also been indirectly addressed by Whetherill *et al.* (2002). They assert that the construction organisation's only sustainable advantage lies in its capability to learn faster than its competitors and the rate of change imposed by the external environment, and that there is a need to 'integrate learning within day-to-day work processes'.

The strategy of the 'live' capture proposed by Kamara *et al.* (2003) adopts the aforementioned combined soft and hard approaches, which attempts to address the cross-organisational knowledge transfer issues (through collaborative learning and learning histories in particular) and to facilitate 'live' capture and reuse of knowledge (through Web-based technology) respectively. The 'live' capture and reuse of project knowledge will (Kamara *et al.*, 2003):

- *Facilitate the reuse of collective learning on a project by individual firms and teams involved in its delivery.* In addition, other project teams can use the learning captured from previous/similar projects to deal with problems; reflection on previous learning can also trigger innovative thinking (to think about issues that might be relevant to their project).
- *Provide knowledge that can be utilised at the operation and maintenance stages of the facility's life cycle.*
- *Help to solve the aforementioned cross-organisational knowledge transfer problems.* The 'live' methodology involves the members of the supply chain in a collaborative effort to capture learning in tandem with project implementation, irrespective of the contract type used to procure the project from the basis for both ongoing and post-project evaluation.
- *Benefit the client organisations with enriched knowledge about the development and construction of their facilities.* This will contribute to the effective management of facilities and the commissioning of other projects. In the longer term, clients will benefit from the increased certainty with which construction firms can predict project outcomes.
- *Benefit the construction industry as a whole.* The construction supply chains will benefit, in both the short- and long-term, through the

shared experiences that are captured as part of the learning on key events (e.g. problems, breakthroughs, change orders, etc.). In the short-term, project teams would be able to better manage the subsequent phases of a project through the capture and transfer of learning from a previous phase. In the long-term, it will increase their capacity to better plan future projects and improve their ability to collaborate with other organisations. Project staff and students of project/construction management and the institutions providing such courses/training will also benefit through the use of captured project knowledge as case study material.

Other potential benefits identified include:

- *Prevention of knowledge loss due to time lapse in capturing the knowledge.* Ebbinghaus's (1885) study reveals that the percentage of human memory retained on a set of data depletes over time. Corresponding to this, the probability of forgetting an event of everyday live (which may include the learning event where new learning is created) is increasing as time elapses (Linton, 1975). Therefore, by facilitating the capture of the knowledge as soon as it is created or identified, 'live' capture of knowledge helps to reduce the loss of knowledge or important insights due to time lapse and to ensure the completeness of knowledge captured.
- *Maximisation of the value of reusing the knowledge captured through 'live' reuse.* 'Live' capture and 'live' reuse of knowledge are interconnected. The true benefit of capturing knowledge comes only when the knowledge is being used (McGee, 2004), particularly if the knowledge is being reused 'live' after it has been captured. Siemieniuch and Sinclair (1999) assert that knowledge can become obsolete and the value attached to the knowledge deteriorates as time passes and the competitive environment for its reuse changes. Some knowledge (used synonymously with data in this context) is required in real time so that effective responses can be deployed at the right time, thereby avoiding mishaps and more importantly seizing opportunities before it is too late (McGee, 2004). McGee (2004) argues that as time passes after an event the possible responses to the event narrow, depicted by the area of triangle in Figure 2.3. This shows that the potential value of 'live' reuse of knowledge in an event may as well narrow and diminish towards the end of the event where the knowledge can be reused, depicted by the area of the triangle in Figure 2.4. This is particularly obvious when the benefit accrued through reusing the knowledge is time-related (e.g. when the knowledge can lead to a saving of £x/day).
- *Help to seize every knowledge reuse opportunity.* Another unique situation is that the knowledge captured may have limited number of events for reuse and hence has to be disseminated for reuse as soon as

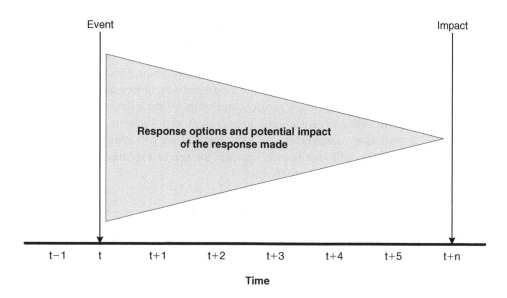

Figure 2.3   As time passes after an event, the possible responses to the event narrow, depicted by the area of the triangle (*Source*: Adapted from McGee, 2004)

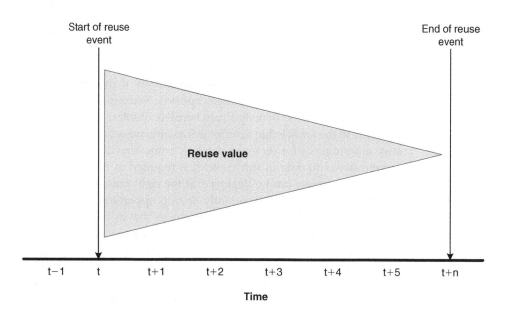

Figure 2.4   As time passes, the potential value accrued through reusing knowledge reduces, depicted by the area of the triangle

possible (i.e. 'live') before such events diminish, as in the Three Mile Island (TMI) incident mentioned by McGee (2004). In the instance, a lesson had been learned that the relief valve used in a nuclear reactor built by Company X had failed to function properly in one occasion and further action was needed. The immediate event for reusing the lesson learned was at the TMI nuclear reactor where another relief valve built by Company X was being used. However, the lesson learned from reactor A was not transferred to the TMI operator and hence was not being 'reused'. No action was taken by TMI operator which had subsequently led to the TMI incident when the valve failed. All other similar type of relief valves would have been replaced after that, but the single most crucial event for reusing the lesson learned had been missed. Similarly, in the context of the construction industry, it is possible for a construction organisation to have similar types of project running roughly parallel, such as the PFI projects. Specific knowledge captured from a particular type of project may only be valuable and reusable in similar projects. In this instance, the 'live' capture and reuse methodology can help to ensure that the specific knowledge created or identified from one courthouse project is made available for sharing and reuse in another courthouse project (or just another courthouse in the same project) so that the triggers for reuse are not missed.

In summary, the result of the review depicts that 'live' capture and reuse of project methodology is crucial in addressing the aforementioned shortcomings of current practice, to better manage project knowledge and to enable the benefits of knowledge captured to be fully exploited. The 'live' capture and reuse of project knowledge methodology facilitates the capture of project knowledge as soon as the knowledge is created or identified (i.e. 'live') to avoid persistent knowledge loss problem of current practice due to time lapse and other constraints. The importance of 'live' methodology is further strengthened by the reduction of knowledge loss facilitated which helps to ensure that a more complete set of project knowledge is captured from construction project, and which allows the knowledge to be reused 'live' to reap the most from the knowledge.

Having established the importance of the 'live' capture and reuse of project knowledge in construction, the next chapter explores the potential types of reusable project knowledge in construction, the learning situations where most of the new learning is created, the current practice for the capture of knowledge focusing on the capability to facilitate 'live' capture of project knowledge and the soft (organisational, cultural and human) issues that affect the knowledge capture and reuse.

3 Reusable Project Knowledge – Generation and Capture

This chapter explores the various types of reusable project knowledge in construction, the learning situations in which most of the new knowledge is generated, the current practice for the capture of new knowledge focusing on the capability to facilitate 'live' capture of project knowledge and the soft (organisational, cultural and human) issues that affect the knowledge capture and reuse.

3.1 Reusable project knowledge

Reusable project knowledge is defined as project knowledge which may be reused in subsequent stages of the same project, or other projects with necessary adaptation to avoid the reinvention of the same knowledge, prevent recurrence of the same mistakes and for continuous improvement. Clearly, identifying the various types of reusable knowledge available before attempting to capture all knowledge is more likely to produce a successful result. Existing literature reveals that there are various types of knowledge identified in the context of knowledge management (KM) which are different both in terms of scope and nature (Maier, 2002). The types of knowledge listed in Table 3.1 are by no means exhaustive but illustrate the varieties available. For the ease of discussion, the various types of knowledge are grouped into construction-domain specific and generic perspectives as shown in Table 3.1.

For the generic perspective, the foremost tacit–explicit distinction drawn by Nonaka and Takeuchi (1995) and Polanyi (1958) is found to be at a level too high for the purpose of identifying reusable project knowledge as most (if not all) of the knowledge are covered under the wide umbrella of tacit and explicit knowledge. The same applies to the distinctions put forward by Bhatt (2001) and Blacker *et al.* (1993). The types of knowledge identified by Ruggles (1997b) and KPMG (1998) cover all the potential areas for knowledge capture in organisations. However, the scope might be too broad for the purpose of capturing reusable project knowledge as discussed in this book. For instance, 'cultural knowledge' is knowledge to be managed at an organisational level rather than

Table 3.1   Classifications of knowledge

Authors	Classification of knowledge	
(a) Generic perspective		
Nonaka and Takeuchi (1995); Polanyi (1958)	• Tacit knowledge	• Explicit knowledge
Bhatt (2001)	• Foreground knowledge	• Background knowledge
Blacker et al. (1993)[1]	• Embrained knowledge • Embodied knowledge • Encultured knowledge	• Embedded knowledge • Encoded knowledge
Rollett (2003: p. 36)	• Core knowledge • Innovative knowledge	• Advanced knowledge
Ruggles (1997b)	• Process • Factual	• Cultural • Catalogue
KPMG (1998)	• Methods and processes • Company's own markets • Company's own products and services	• Regulatory environments • Customers • Competitors • Employee skills
(b) Construction-domain specific perspective		
McLoughlin et al. (2000)	• Know-how • Know where/when	• Know why • Know what
Whetherill et al. (2002)	• Project • Organisational	• Domain
Robinson et al. (2001)	• Process • People	• Product
Kamara et al. (2002b)	• Organisational processes and procedures • Client's business • How to predict outcomes, manage teams, focus on clients and motivate others	• Technical/domain knowledge • Know-who

[1] Embrained knowledge relates to the conceptual skills and cognitive abilities of individuals; embodied knowledge is action-oriented and is rooted in specific contexts; encultured knowledge refers to the process of achieving shared understanding; embedded knowledge is knowledge which resides in systematic routines and encoded knowledge is information conveyed by signs and symbols.

at a project level. A scope which is too wide could result in unnecessary knowledge overload and affect the knowledge capture and reuse processes.

For the construction-domain specific perspective, McLoughlin *et al.*'s (2000) four types of knowing have added insights into the scope of knowledge to be managed in long-term engineering projects but they are less helpful in identifying the exact types of reusable project knowledge. Whetherill *et al.* (2002) note that knowledge in this perspective (i.e. organisational, domain and project knowledge) are strongly interlinked in that any amendment introduced to one category is very likely to have a critical impact on the others. Robinson *et al.*'s (2001) findings reveal that the knowledge within the construction domain can be grouped into the three context-based factors: process, product and people. The three context-based factors relate to the issues of what is produced (products), how it is produced (processes) and by whom (people) (Robinson *et al.*, 2001). The various types of knowledge identified by Kamara *et al.* (2002) serve as a useful guide to the various types of knowledge that exist, but it must be noted that it is not solely based on the perspective of a construction organisation. The exact types of reusable project knowledge to be captured in construction have therefore been identified from detailed case studies which are described in detail by Tan *et al.* (2004).

3.1.1 *Types of reusable project knowledge*

Through the case studies, a wide spectrum of reusable project knowledge was identified. Different companies used different terms to describe similar types of knowledge. The types of reusable project knowledge identified were therefore aligned and grouped into categories. It is possible to break some of the categories down into different types of reusable project knowledge. This list is by no means exhaustive but represents the main categories identified:

- *Process knowledge*: This is the knowledge pertaining to the execution of various stages of a construction project. The types of reusable project knowledge that belong to this category include design, tendering and estimating, planning, construction methods and techniques and operation and maintenance knowledge. These knowledge types are sometimes captured in the form of standard procedures (e.g. design manual for design knowledge) but mostly remain tacit.
- *Knowledge about clients*: This covers the knowledge about clients' specific requirements, their internal procedures and business. This knowledge may exist in the form of standard procedures compiled based on the experience of dealing with clients. It may also remain tacit and is usually shared through interactions between people.
- *Costing knowledge*: This knowledge is about the costs of alternative forms of construction and the whole life cost (WLC) of a facility. This knowledge may remain tacit (in the heads of estimators), or be explicated and captured in custom-designed software.
- *Knowledge about legal and statutory requirements*: Regulatory requirements change over time. This knowledge covers the requirements and responsibilities imposed by regulations, codes of practice, and the best practice to address these requirements. This knowledge is available through subscription to the relevant Web services and in the form of CDs. It may also be held in the heads of people through experience or attending specific courses.

- *Knowledge about reusable details*: Reusable details comprise standard design details, specifications and method statements. These details may be reused with adaptations. They help to avoid recreating similar details from scratch and also lead to time and cost savings.
- *Knowledge of best practices and lessons learned*: These are the proven ways of working that contribute to the success of projects, and the mistakes made that must be avoided in future projects. This knowledge is often explicated and compiled as best practice guides and codes of practice.
- *Knowledge of performance of suppliers*: The suppliers referred to are consultants, contractors, subcontractors, material suppliers and others who contribute services or goods to a project. The capture of this knowledge facilitates better selection of suppliers for future projects. This knowledge is explicit in nature. It is often captured in a custom-designed database which is accessible through intranet.
- *Knowledge of who knows what*: This is the knowledge of the skills, experience and expertise of each member of staff. It helps to locate the right people with the right knowledge for the sharing of knowledge, particularly the tacit knowledge which is difficult to codify. This knowledge is captured in organisation's staff profile or skills yellow pages.
- *Other types of knowledge*: This knowledge category includes key knowledge about competitors, risk management, key performance indicators and other sector-specific knowledge (e.g. knowledge about flood control and management of water networks).

The various categories of reusable project knowledge identified can also be classified according to project types (e.g. hospital project) and procurement routes (e.g. PFI or Private Finance Initiative) as appropriate. This can help in the location of the relevant reusable project knowledge that is captured from projects that use a particular type of procurement route or fall under a particular project type. For the details of the different categories of reusable project knowledge, see Appendix B.

3.1.2 Characteristics of reusable project knowledge

The nature and characteristics of reusable project knowledge identified from the case studies were consistent with its proposed definition (i.e. it can be adapted for reuse in different situations, and that it can lead to further improvement and innovation). Further insights about reusable project knowledge obtained from the case studies include:

- It is borne out of a set of particular circumstances that exist on a recurrent basis.
- It is adaptable (i.e. the new application may not be identical but the knowledge is capable of adaptation to make it work in the new context).

- The reuse of project knowledge is not limited to the same project or other similar projects, but also in other departments.
- It is capable of being transferred across sectors for reuse (e.g. from construction to the manufacturing sector).
- It is the amalgamation of an industry's and a company's previous knowledge complemented by new research findings, new ways of working and new ideas, which ultimately leads to innovation and improved best practice.

The finding that reusable project knowledge is borne out of a set of particular circumstances that exist on a recurring basis has indirectly proven the existence of learning situations (see Section 3.2 for details) within which knowledge is created. It also highlights the relationship between reusable project knowledge and the various stages of the construction project, as well as the possibility of identifying and capturing the reusable project knowledge based on project stages. Existing process models that outline the project stages which can be used for the purpose includes the RIBA Plan of Work, and Process Protocol Map developed by Kagioglou et al. (1998).

Table 3.2 shows the possibility and attempts to identify and capture reusable project knowledge based on the Process Protocol's phases. The four broad Process Protocol phases are pre-project, pre-construction, construction and post-construction. Different types of reusable project knowledge can be captured at the project reviews conducted at different project stages. However, some knowledge types may be captured throughout the duration of a project.

The types of reusable project knowledge identified by the case study companies were found to be centred on their core activities. For instance, the key types of reusable project knowledge identified by a design consultancy are design knowledge, regulatory requirements knowledge and knowledge about the client's specific requirements which have a bearing on the design. For a management consultancy, its reusable project knowledge is wide ranging and covers the management knowledge of the various project processes. In the case of a water company, technical and engineering knowledge pertaining to the management of its water network were regarded as important reusable project knowledge. Therefore, although the list of reusable project knowledge presented is reasonably comprehensive, it is conceivable that there will be other specific types of reusable project knowledge which may be considered important by other companies.

Further analysis revealed that the reusable project knowledge identified often exist as a mix of tacit and explicit knowledge, rather than as distinctive tacit or explicit knowledge alone. For instance, two companies had externalised part of their design knowledge and knowledge about best practices into the form of a design manual and technical notes

Table 3.2   Types of reusable project knowledge identified and the Process Protocol stages at which the knowledge can be captured

Reusable project knowledge	Process Protocol phases			
	Phase 1	Phase 2	Phase 3	Phase 4
1. Process knowledge				
• Briefing	•	•		
• Design	•	•		
• Tendering and estimating	•	•		
• Planning		•		
• Construction and buildability			•	
• Operation and maintenance			•	•
2. Knowledge about client				
• Clients' requirements	•	•		
• Client organisations' internal procedures	•	•	•	
• Background knowledge about client's business	•	•		
3. Costing knowledge				
• Cost of alternative forms of construction			•	
• WLC			•	•
4. Knowledge of legal and statutory requirements				
• Health and safety			•	•
• Changes in regulatory requirements	•	•	•	•
• Contract	•	•	•	•
5. Knowledge of reusable details				
• Standard design details	•	•	•	
• Specifications	•	•	•	
• Method statements	•	•	•	
6. Knowledge of best practices and lessons learned	•	•	•	•
7. Knowledge of performance of suppliers			•	•
8. Knowledge of who knows what				•
9. Other types of knowledge				
• Risk management	•	•		•
• Team working	•	•	•	•
• Project management	•	•	•	•

respectively. However, both companies acknowledged that there is still some knowledge which is difficult to externalise and hence remains tacit (in the heads of people). The findings suggest that in addition to the tacit and explicit dimensions, there is an additional dimension of knowledge (i.e. the tacit knowledge which can be made explicit but has not yet been converted). This is depicted in Figure 3.1.

As explicit knowledge is comparatively easier to be transferred and shared for reuse, the methodology designed for live capture and reuse of project knowledge is intended to convert tacit knowledge into explicit knowledge as far as possible. For the remaining tacit knowledge which is

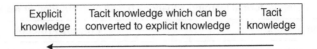

Figure 3.1   **Three dimensions of knowledge**

really difficult to convert, mechanisms should be provided to connect the people with a particular type of knowledge and the people who need the knowledge for the sharing of the knowledge. These findings are reflected in the design of the methodology for the 'live' capture and reuse of reusable project, as discussed in Chapter 5.

3.2   Learning situations

In discussing the 'live' methodology for the capture and reuse of project knowledge, Kamara *et al.* (2003) introduced the notion of 'learning events' which are synonymous to learning situations. Learning situations are a range of circumstances that emerge during the course of the project where new knowledge can be generated and captured. They can be critical events and also normal day-to-day operations (Kamara *et al.*, 2003). Identification of the various learning events in advance is helpful in the development of a methodology for the capture of reusable project knowledge. There are at least two categories of learning situations, namely formal and ad hoc learning situations.

3.2.1   *Formal learning situations*

Formal learning situations are routine events, such as the weekly site meetings, project reviews conducted at the end of each of the project stages (e.g. feasibility study, full conceptual design and construction stages) or at predetermined intervals and post project reviews (PPRs). Among the examples given, PPRs have been identified as the most common approach for the capture of learning from construction projects (Orange *et al.*, 1999). Formal learning situations can also be identified through the RIBA Plan of Work and Process Protocol depicted in Table 3.3. This illustrates the various project stages/phases, which can offer the potential for formal learning situations.

3.2.2   *Ad hoc learning situations*

Ad hoc learning situations are the non-routine special learning situations such as problems and unforeseen circumstances encountered, which have a bearing on the performance of the project. Ad hoc learning situations such as problems may lead to the capture of 'lessons learned', whereas the solutions formulated to resolve the problems may contribute towards

Table 3.3 Potential formal learning situations based on RIBA Plan of Work and Process Protocol

RIBA Plan of Work	Process Protocol
• Stage A: Appraisal	• Phase 0: Demonstrating the need
• Stage B: Strategic brief	• Phase 1: Conception of need
• Stage C: Outline proposals	• Phase 2: Outline feasibility
• Stage D: Detailed proposals	• Phase 3: Substantive feasibility study and outline financial authority
• Stage E: Final proposals	• Phase 4: Outline conceptual design
• Stage F: Production information	• Phase 5: Full conceptual design
• Stage G: Tender documentation	• Phase 6: Coordinated design, procurement and full financial authority
• Stage H: Tender action	• Phase 7: Production information
• Stage J: Mobilisation	• Phase 8: Construction
• Stage K: Construction to practical completion	• Phase 9: Operation and maintenance
• Stage L: After practical completion	

the creation of 'best practice'. It is therefore believed that a great proportion of new learning from construction projects is created in the learning situations. Hence, identifying the various learning situations is crucial for the capture of reusable project knowledge in construction. However, most of the construction companies involved in the case studies found it difficult to identify specific learning situations where new knowledge was created and captured. This is because they asserted that every decision-making process and 'every situation that emerges during the course of the project' has the potential to be a learning situation. As a result, people may not realise that a particular situation is a learning situation when they are facing one. This is due to the fact that their attention is often concentrated on resolving the issues that have arisen in the learning situations. Some of the learning situations identified are:

- New project location/market: When a company has a new project in a new area or another country with a different set of regulations and local issues, this local knowledge must be captured.
- New type of project: A new type of project often has different and specific requirements or characteristics which necessitate the learning of new technical or management practices, construction methods and the use of new technology.
- Change of end-user/client: There are circumstances where the ownership of a building which is originally designed for an end-user/client based on his/her specific requirements may be transferred to another party during the course of the contract for certain reasons. As a result,

the building will have a new end-user or a new client and changes in design may be required in order to address the new requirements to suit the new purpose of use.

- Problem in the supply of major building fabric/material: The desired building fabric or materials in the design, particularly those very specific items such as special types of space frame from a particular manufacturer, can become unavailable during the course of the project in some instances. This will impose new problems to the project team.
- Undiscovered condition of the project: Using a refurbishment project as an example, during the course of the project some defects and faults which were not discovered initially may be identified. This can lead to a new set of problems, extension of project duration, additional work and may even have an impact on the current job nature.
- Change in political climate: A change in government may lead to policy changes that may affect the construction industry.
- Political problems: These could be potential objections to the construction of a new facility (e.g. airport, water treatment plant or road).
- New learning can also be captured based on the project stages as delineated in the Process Protocol and RIBA Plan of Work.

The ad hoc learning situations identified are centred on the management of change (e.g. the change in political climate and clients, involvement in new types of project in another country), problems (e.g. undiscovered condition of project and the supply of major building materials) and the adaptation required in response to the changes and problems. Ad hoc learning situations are very similar to the 'triggers of knowledge production' identified by Egbu et al. (2001) where knowledge is being produced and captured. Egbu et al. (2001) grouped the triggers of knowledge production in construction organisations into three categories:

- problem solving
- managing change
- innovation

Problem solving can be regarded as a learning situation since the process involved and lessons learned in solving major problems are reusable project knowledge. For change management, the industry and major clients have recognised the need to identify and manage the change properly (Lazarus and Clifton, 2001). Regarding innovation, Egbu et al. (2001) note that it is the crucial driving force behind knowledge production in the construction industry and is, therefore, the source of new learning from projects. More ad hoc learning situations are identified by identifying the major issues of concern within each of the three categories of triggers for knowledge production (see Appendix C). The learning situations identified from the case studies and existing literature are summarised in Table 3.4.

Table 3.4 Table summarising the ad hoc learning situations/triggers of knowledge production identified from this research

Ad hoc learning situations		
Problem solving	**Managing change**	**Innovation**
(a) Triggers of knowledge production (Egbu *et al.*, 2001)		
• Dealing with complex projects • Managing team member interfaces (e.g. consultant–contractor) • Addressing value engineering issues to deliver the best value • Addressing clients' needs • Identifying the knowledge gap • Dealing with contextual differences • Finding measures to increase company competitiveness • Dealing with lack of (design) information • Addressing the need to improve the quality of product/service • Addressing the need to improve efficiency • Addressing the need to recruit skilled people and retain them • Dealing with challenging site logistics • Dealing and coping with incompetent consultants	• Managing changes to the project • Managing organisational change • Addressing the need to comply with standards (Quality Assurance, Health and Safety, etc.) • Addressing the changes to statutory regulations, technical standards • Dealing with contractual arrangements new to the respondent • Being enabled to make design choices • Working with new sub-contractors • Being assigned to a new role • Addressing the need to establish a data transformation system for the whole project team • Addressing the need to create a 'database' • Coping with Government initiatives (e.g. PFI, partnering)	• Using new, innovative building materials, systems, services • Coping with the uniqueness of projects • Dealing with the need and willingness to be 'ahead of the game', 'move the market' • Addressing the pressure and need to innovate ('look at new ways of doing things')
(b) Ad hoc learning situations identified from the case studies and existing literature		
• Supply of major building fabric/material • Undiscovered condition of project • Site condition change • Inflation or relative price rise • Difficulties with contractors • Termination and default • Projects behind schedule • Claims and disputes • Budget related issues • Human resource issues • Political problems	• New project location/ market • New type of project • Change of end-user/client • Change in project scope • Professional errors and omissions • Design change • Change in client's requirements • Change in construction method, etc. proposed by contractor • Changing market requirements • Change in political climate	• Fundamental and invasive technology improvements • New process that has benefits to the company • New approach to providing services to customers/clients • New procedures for obtaining goods/services • New product that provides competitive advantage for the company • New external relations, for example partnering and joint ventures • New administrative policy, for example incentive schemes and bonuses

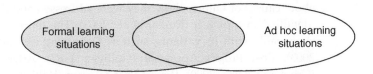

Figure 3.2   Relationship between formal and ad hoc learning situations

Further analysis reveals that formal learning situation often involves a group of people for the capture of learning (e.g. project meetings), whereas the ad hoc learning situation may involve only individuals. However, formal and ad hoc learning situations are not mutually exclusive as the issues that emerge from ad hoc learning situations may be raised for further discussion in the meetings and reviews with the aim of obtaining better solutions. The design of a knowledge capture and reuse methodology must be capable of capturing knowledge from a group of people and individuals so that the knowledge created in both types of learning situations will not be overlooked. The relationship between formal and ad hoc learning situations is depicted in Figure 3.2, which shows a certain degree of overlap.

3.3   Current practice on capture and reuse of project knowledge

KM practice depends significantly on the use of various KM tools to perform the KM sub-processes (e.g. knowledge capture, sharing and reuse). KM tools can be either IT-based or non-IT based. There is confusion over the definitions of KM tools as most authors infer KM tools to mean Information technology (IT) tools (Al-Ghassani, 2002). However, as Ruggles (1997b) points out, not all KM tools are IT-based. Everyday tools such as papers, pens and photos can be utilised to support KM. To avoid confusion, Al-Ghassani (2002) proposed that the terms 'KM Techniques' and 'KM Technologies' to be used to replace 'Non-IT tools' and 'IT tools' respectively. This distinction is used for the purpose of the research.

Here, there is no attempt to cover each of the KM techniques and KM technologies in detail because a wide range of them are available for different KM sub-processes (Ruggles, 1997a; Jackson, 1998; Laudon and Laudon, 2000; Wensley, 2000; Gallupe, 2001; Rezgui, 2001; Al-Ghassani, 2002; Tsui, 2002). The following section discusses only the most relevant and important KM techniques and technologies for the capture and reuse of project knowledge.

3.3.1   Post project reviews

PPRs are debriefing sessions used to highlight lessons learned during the course of a project. These are important to capture knowledge about causes of failures, how they were addressed and the best practices identified in a project (Ruikar *et al.*, 2003). The term is also used interchangeably with its variants, such as Debriefing (Schindler and Eppler, 2003), Post Project Appraisal (Gulliver, 1987), After Action Review (Cross and Baird, 2000), Project Post-Mortem (Williams *et al.*, 2001), Post Implementation Evaluation (Kartam, 1996), Project Audit, Project Close-out and Post Completion Review. PPR is the most common approach used in construction industry to capture the learning from projects (Orange *et al.*, 1999).

Facilitation of knowledge capture and reuse

The importance of PPR in knowledge capture and reuse is related to the temporary nature of project teams. After a project ends, the team is dissolved and it is hard to access the learning from the project (Disterer, 2002). Therefore, if planned in advance, the PPR is the last chance for an organisation to capture learning from the project so that it can be transferred to other projects.

PPRs are also tied to collective and group learning where they are seen as a means to synergise the learning of individuals. Kerth (2000) points out that 'there are many pieces to the whole story of the project, and each individual on the project knows his piece of [the] story', and until everyone on the team joins together and collectively tell the story, the learning is likely to be minimal. The collective telling of the story facilitated by PPR illuminates pieces of the project that no one can see by themselves (Kerth, 2000). Therefore, by gathering the whole team together PPRs aid knowledge capture and reuse across the team with the potential for distributing the learning across organisations.

Disterer (2002) argues that the documented 'lessons learned' from PPR can play a significant role in externalising, storing and sharing tacit knowledge. This is because the 'lessons learned' document covers the full and detailed description of the problems and how they were solved taking into consideration technical issues, organisational issues and social situations (Disterer, 2002). Therefore, tacit knowledge can be acquired through an understanding of the process and the underlying mental models and insights.

Issues and capability to support 'live' capture and reuse of project knowledge

As PPR is conducted at the end of a project, it certainly does not facilitate the 'live' capture and reuse of project knowledge in construction. Other than this, it is also undermined by the following issues:

- *Time constraint*: This is a critical issue for PPR. There is time pressure towards the end of a project as the team strives to meet the completion deadline and new tasks already await the dissolving team (Schindler and Eppler, 2003). With personnel often transferred to other projects and others soliciting new work, project team members can only be identified with huge effort (Disterer, 2002) and people may not want to dedicate time to review past issues (Kartam, 1996). Therefore, PPRs are sometimes treated as a burden to be rushed through so that attention can move on to more pressing matters (Kartam, 1996).

- *Reluctance to share mistakes*: Apart from the loss of important information or insights due to time lapse in capturing the learning (Kamara *et al.*, 2003), there is also insufficient willingness for learning from mistakes of the person involved (Schindler and Eppler, 2003) where mistakes are deliberately forgotten and not to be disclosed (Kartam, 1996).
- *Objectivity*: The objectivity of findings and opinions of PPR are questionable in some cases. Shapiro (1999) argues that as PPRs are undertaken retrospectively, they are susceptible to 'the characteristics partial and selective memory recall' by managers who, after the event, are rarely neutral or objective.
- *Lack of a format for representing knowledge*: Kartam (1996) argues that the most serious shortcoming of PPR is a failure to uniformly document lessons learned in a manner useful to others in the future. He further points out that the lack of such format renders the retrieval of lessons learned for use in future work difficult. Corresponding to this, Schindler and Eppler (2003) identify that the compiled result of a PPR might be described too generically and may not be visualised as necessary – this prevents reuse due to the lack of context. Also, it may be archived in a way that others have difficulties in retrieving the captured knowledge. Moreover, the result of PPRs may not be accepted although they are well documented and easy to retrieve due to the so-called 'not-invented here syndrome' (Schindler and Eppler, 2003).

3.3.2   Communities of Practice

The term Communities of Practice (CoPs) is often used interchangeably with communities of knowing, knowledge communities, knowledge networks, learning communities, communities of interest and thematic groups. Wenger *et al.* (2002) define CoPs as 'groups of people who share a concern, a set of problems, or a passion about a topic, and who deepen their knowledge and expertise in this area by interacting on an ongoing basis'. There are a number of different types of CoPs based on dichotomised categorisation (informal/formal, organic/structure, natural/engineered), or as a range of types (informal, supported and structured) as identified by Saint-Onge and Wallace (2003). Members of CoPs possess different skill sets, development histories and experiences but they have commonly shared goals that they are working together to achieve (Ruggles, 1997a).

Facilitation of knowledge capture and reuse

The significance of CoPs towards KM is generalised by Saint-Onge and Wallace (2003) as providing a platform for their members to pool their expertise, experience, and ideas, and to find solutions. Other importance of CoPs includes:

- CoPs can provide better access to knowledge bases within their collaborative space and that located externally through affiliated professional organisation's Internet site or personal sources (Saint-Onge and Wallace, 2003). Indirectly, CoPs can help to group the knowledge of individuals within the community into collective knowledge resources for the benefits of their members.
- CoPs are important in the sharing of tacit knowledge which is typically based on experience (Newell et al., 2002). This is because these communities share 'a common experience of practice' and their members have developed a set of shared meanings deriving from their common experience (Newell et al., 2002). Therefore, the basic assumptions and contextual features which are closely related to tacit knowledge and provide insight into the knowledge shared do not have to be explained and are readily understood.
- In CoPs, new knowledge may be created through an incremental improvement of an idea that results from the synthesis of community members' contributions in a brainstorming session. This can also be achieved through collaborative problem solving facilitated over a period of time which is supported by external expertise and access to additional resources (Saint-Onge and Wallace, 2003).
- Best practice guides can also be produced from the result of CoPs members' interactions and discussions. The guide, which is the members' externalised tacit knowledge, can be channelled back to everyone in the organisation and expand the organisation's knowledge base.
- *Timeliness of knowledge*: CoPs, particularly those aided by IT, may allow the user to obtain the knowledge required quicker than other sources (Ardichvili et al., 2003).

Issues and capability to support 'live' capture and reuse of project knowledge

Information and communication technology (ICT) can be used to link geographically dispersed CoPs together to facilitate communication between these CoPs (Ardichvili et al., 2003). From here, it is possible that online CoPs can be one of the KM tools capable of supporting 'live' sharing and reuse of knowledge in construction. However, in terms of 'live' capture, its role is found less significant unless a mechanism is developed for the capture of knowledge from members of CoPs once knowledge is created or identified.

Furthermore, the efficacy of knowledge sharing and transfer in CoPs is deeply influenced by the organisational culture and trust. Newell *et al.* (2002) point out that knowledge-sharing in CoPs is actually facilitated by the norms of reciprocity – 'you help me and I will help you' – and the level of trust generated amongst the community. If the supportive culture does not exist, cultural change is necessary to create a supporting behaviour around knowledge sharing in CoPs.

3.3.3 Training

Training programmes are organised for employees changing job description or being promoted, and to enhance their skills and knowledge. In the context of KM, training can be considered inseparable to 'learning' and 'skill', and is important for the implementation of KM systems (Harman and Brelade, 2000). Training can be broadly categorised as conventional training and the training aided by ICT. Conventional training is instructor-led and mainly involves face-to-face interactions. ICT-based is also referred to as online training, net-based training (Gotschall, 2000) or computer-based training (Zahm, 2000). This form of training is usually delivered via CD-ROM or through downloads from the Internet or an intranet (Gotschall, 2000; Zahm, 2000). Maier (2002) indicates that ILOI's (1997) and Bullinger *et al.*'s (1997) studies reveal that 83% of the organisations reported personnel training and education as the most important KM instrument for experiences, and as the most frequently used instrument for knowledge acquisition respectively.

Facilitation of knowledge capture and reuse

The role of training in KM includes the following:

- Training can aid individual and organisational learning. In organisations, only people are said to be able to learn and organisations ultimately learn from individuals (Hwang, 2003). Through training, however, an organisation's collective knowledge can be transferred back to individuals. Training may also help to transfer tacit knowledge through the face-to-face interactions of people involved during the events.
- Training can aid personal KM. It can help the development of personal KM capacity, that is the ability to evaluate, learn, structure, share and use knowledge, using the organisation's KM systems (Vorbeck and Finke, 2001). The importance of this type of training can be viewed from Hwang's (2003) contention that 'although there may be a wonderful KM system built, knowledge cannot be directed at sustaining profitability if people do not have the skill or ability to use knowledge creatively'. Furthermore, training can also be designed to help create

a supportive KM culture by nurturing the open-mindedness as well as the self-motivation among staff towards capturing and sharing knowledge.

Issues and capability to support 'live' capture and reuse of project knowledge

Explicit knowledge can be disseminated in the form of handouts and tacit knowledge can be shared through face-to-face interactions during the training sessions. However, there is little evidence of the capability of training to support the 'live' capture and reuse of project knowledge. This is possibly because it is very difficult for training on certain topics to be conducted immediately for the capture and sharing of the relevant knowledge. The role of training is more significant in terms of the sharing than the capture of project knowledge. However, there is scope for training to be used to sensitise project participants about the potential for 'live' capture of knowledge, and teach them how to recognise and capitalise on learning situations.

3.3.4   Recruitment

Recruitment is the process of finding new people to join a company, and is usually an effective means of bringing new knowledge into an organisation.

Facilitation of knowledge capture and reuse

Recruitment is regarded as one of the easiest ways to acquire or capture new knowledge especially when an organisation is engaged in a new project sector (Tan, 2002: p. 84). Current literature also suggests that recruitment should be geared towards getting new people to fill existing and future anticipated knowledge and skills gaps (Harman and Brelade, 2000).

Recruitment adds new knowledge and expands the organisational knowledge base, and allows other members of the organisation to learn from the recruited member (Ruikar et al., 2003). This approach might prove successful in many situations as creditability is often higher for external experts and an organisation's experts might be more willing to accept and reuse ideas from outside the organisation than from within (Robertson, 1999). Furthermore, the introduction of new recruits into organisations may ask for knowledge explication that stimulates the exploration necessary for innovative activities and for the creation of new knowledge (Levina, 1999). Some organisations also attempt to codify the recruited person's knowledge that is of critical importance to their business (Ruikar et al., 2003).

Issues and capability to support 'live' capture and reuse of project knowledge

Maier (2002) identifies the following difficulties in recruiting experts as a means to acquire new knowledge:

- Experts are scarce and it is therefore difficult to recruit and retain them (Maier, 2002). This problem is magnified by competitors' constant attempts to entice knowledge workers from their rivals (Robertson, 1999).
- It is difficult to integrate experts into the organisation's knowledge networks, culture and processes so that core competencies can be built-up.
- It is difficult to assess the capability of experts. The assessment process can also be very costly (Harman and Brelade, 2000).

According to Ruikar *et al.* (2003), recruitment is a measure to acquire new knowledge to expand an organisation's knowledge base, rather than a 'live' knowledge capture and reuse technique. Furthermore, the issues of recruitment identified by Maier (2002) and the fact that recruitment is a lengthy time-consuming process also indicates that it cannot be considered a 'live' knowledge capture technique.

3.3.5   Face-to-face interaction

This is the oldest (Ribes *et al.*, 1981), most fundamental yet powerful form of knowledge capture and sharing practice in organisations (Ruikar *et al.*, 2003). It is reliant on people meeting 'face-to-face'. Therefore, despite the fact that there are a plethora of KM techniques and technologies available, the human channel via face-to-face interaction is still regarded as the most effective way of knowledge sharing (Davenport *et al.*, 1997).

Facilitation of knowledge capture and reuse

Face-to-face interaction is important in the transfer of tacit knowledge, the richness of which is difficult to document (Hansen *et al.*, 1999; Ladd and Heminger, 2002; Engström, 2003). Hansen *et al.* (1999) highlight a case where Xerox once attempted to embed the know-how of its service and repair technicians into an expert system. However, Xerox's attempt failed as technicians learned from others by sharing stories about how they had fixed the machines, and the expert system could not replicate the tacit knowledge exchanged in the face-to-face interactions.

In a project environment, face-to-face interactions within and between project teams have also been identified as a central feature of resolving issues and generating new ideas (Marshall and Sapsed, 2000). Lang (2001) asserts that face-to-face interaction also helps to create social ties and tacit shared understandings which give rise to collective sense-making. This in turn leads to emergent consensus as to what is valid knowledge and the creation of new knowledge.

Issues and capability to support 'live' capture and reuse of project knowledge

There is evidence showing that face-to-face interaction is pertinent to 'live' knowledge capture and reuse. Koskinen *et al.* (2003) state that face-to-face interactions enhance the use of tacit knowledge in engineering projects due to its 'capacity for immediate feedback'. This allows understanding to be checked and any misinterpretations corrected instantly. This method also allows simultaneous interpretation of multiple cues, including body language, facial expressions and the tone of voice, which convey knowledge beyond spoken knowledge (Koskinen *et al.*, 2003), and facilitates 'live' capture of knowledge.

However, the efficacy of 'live' knowledge capture and reuse through face-to-face interactions is influenced by the following factors:

- *Geographical distance*: Face-to-face interaction is not always possible as in the case of multinational organisations. In addition, the fact that a project team consists of personnel from different organisations further compounds the problem.
- An intimate relationship between the source and recipient to remove the barriers of knowledge transfer (Hall *et al.*, 2000).
- The source and scope of knowledge available is restricted to people around the community who can be met with ease when required.
- *Access to knowledge*: Others may not have access to the knowledge shared unless they are also physically involved, or manage to identify someone who is involved in the process.

Therefore, although face-to-face interaction is effective in the sharing of tacit knowledge, the various issues associated with this method suggest that it cannot, on its own, meet the requirements of 'live' capture and reuse of project knowledge.

3.3.6 Mentoring

Mentoring is a practice where new personnel or junior staff are assisted in their work by attaching them to an experienced colleague for a certain amount of time (Al-Ghassani, 2002). Mentoring can be categorised as formal/informal (Chao *et al.*, 1992), and internal/external (Ragins, 1997) based on the formation of the relationship between the mentor and protégé, and the relationship between the mentor and the protégé's organisation. Informal mentorships are not formally recognised and formed by the organisation, whereas formal mentorships are managed and sanctioned by the organisation (Chao *et al.*, 1992). Internal mentors are employed in the same organisation as the protégé. An external mentor is a member of staff from another organisation. Mentorship is primarily focused on 'career and development' (Tabbron, 1997) and aimed at developing necessary transitional competencies when people are given tasks that are new, unfamiliar and fraught with stress and uncertainty (Von Krogh *et al.*, 2000).

Facilitation of knowledge capture and reuse

Kleinman *et al.*'s (2001) research reveals that mentoring helps to expedite and improve learning of the protégés about the context within which the professional works (e.g. on how one's performance affects others and other departments), and broadens their portfolio of skills and abilities by modelling behaviours displayed by the mentor. Under the guidance of the mentor, the protégé can go through a learning process in which he/she creates the tacit and explicit knowledge to accomplish the task, and develops the skills to identify and share this tacit knowledge (Von Krogh *et al.*, 2000).

Mentoring is also crucial in the sharing of tacit knowledge. Engström (2003) argues that effective sharing of tacit knowledge depends on an enabling context, which is likened to a space or an open and trusting relationship between individuals, for the sharing of tacit knowledge. He further argues that mentoring can help in establishing the required enabling context. Apart from this, the face-to-face conversation and interaction associated with mentoring also facilitate the sharing of tacit knowledge (Engström, 2003).

In addition, the protégé is not the sole beneficiary of mentoring. The mentor also benefits from gaining insights into the issues faced by their protégés (Tabbron *et al.*, 1997), and the opportunity to make productive use of his/her knowledge and to learn in new ways (Burke *et al.*, 1994).

Issues and capability to support 'live' capture and reuse of project knowledge

Despite the favourable findings on the significance of mentoring in KM, further study reveals that the effectiveness of mentoring is undermined by the following issues:

- Organisational practices and management decisions (e.g. a decision to promote and transfer a mentor to another department) may decrease the opportunity for interaction between the mentor and protégé (Kram, 1988). This will affect knowledge sharing and transfer through mentoring.
- Cross-gender mentoring, where the mentor and protégé are of opposite gender. This is due to the difficulties for women to find female mentors (Burke *et al.*, 1994: p. 23) and to handle their relationship with

male mentors (Clawson and Kram, 1984), and the likelihood of experiencing greater social distance as well as discomfort with male mentors (Kram, 1988).

- Time constraints (Billett, 2003; Carrillo, 2004). Today's de-layered, lean and complex matrix organisations do not naturally allow the time or offer the right climate and structure to encourage experienced managers or colleagues to voluntarily provide mentoring to protégés (Tabbron et al., 1997).
- Restriction in transfer of knowledge. Mentoring is mainly a one-to-one relationship (Tabbron et al., 1997), thus the transfer of knowledge is more likely to be restricted to that between mentor and protégé.
- Fear of displacement. Experienced members of staff may be concerned about displacement by the protégé whom they have mentored (Billett, 2003).
- Mentor quality. The competence of mentors is attributable to previous experience, depth of and confidence in the knowledge of the work in which they are mentoring, and their ability to develop mentoring skills in a supportive environment (Billett, 2003). Megginson (2000) contends that sufficient training of mentors is crucial to ensure that they have the necessary skills to perform the task.

In terms of the 'live' capture and reuse of knowledge, it is worth mentioning that the knowledge captured 'live' by the mentor may not be shared 'live' with the protégé due to aforementioned issues such as time constraints, distance between mentor and protégé and the perception that the knowledge is not important. In addition, it is unlikely for a protégé to have control over the type of knowledge to be shared by a mentor. Moreover, a protégé's access to the mentors' knowledge is greatly dependant on the availability of the mentor. The issues that undermine the effectiveness of mentoring identified also suggest that mentoring may not be the best tool to facilitate the 'live' capture and reuse of project knowledge.

3.3.7 Succession planning and management

Traditionally, succession planning and management is concerned with the identification of the gaps which are likely to occur in an organisation due to anticipated future changes or known factors such as retirement and reassignment (Harman and Brelade, 2000), the selection of talented employees to fill the gaps to ensure continuity in management practices (Hirsh et al., 1990; Huang, 2001) and development of people (Hirsh et al., 1990).

Facilitation of knowledge capture and reuse

In the context of KM, succession planning and management is extended to 'meeting anticipated knowledge and skills gaps' when there is someone leaving the organisation (Harman and Brelade, 2000), the retention of corporate knowledge (Ministry of the Premier and Cabinet, 1999) and to facilitate the transfer of mainly tacit knowledge between the successor and incumbent (Carrillo, 2004). Findings by Harman and Brelade (2000) and Ministry of Premier and Cabinet (1999) reveal that succession planning and management actually involves the systematic use of other tools and techniques available to achieve its goals. These include the following:

- Training and development (Harman and Brelade, 2000);
- Structured work experience (Harman and Brelade, 2000);
- Formal and informal mentoring programmes (Ministry of the Premier and Cabinet, 1999);
- Formal knowledge transfer forums (Ministry of the Premier and Cabinet, 1999);
- Oral histories/briefings (Ministry of the Premier and Cabinet, 1999; Kransdorff, 1996);
- Electronic systems designed specifically for knowledge transfer (Ministry of the Premier and Cabinet, 1999);
- Imbedded systems to retain innovation (Ministry of the Premier and Cabinet, 1999).

Kransdorff's (1996) finding indicates that the knowledge transferred through this approach are pertaining to corporate culture, management and communication style and the details of recent events which enable one to take over the new tasks with ease and perform the tasks to the standard required quickly.

Issues and capability to support 'live' capture and reuse of project knowledge

Some problems which may undermine the capability of succession planning and management in facilitating 'live' knowledge capture and reuse of project knowledge in construction include:

- Over-concentration on general management skills rather than functional and specialist skills (Hirsh et al., 1990);
- Literature suggests that succession management is related more to the senior rather than lower positions in organisations (Hirsh et al., 1990; Byham, 2002);
- Its reliance on the early identification of potential candidates tends to exclude those who take a later decision to pursue a management career, those who move between employers and those who interrupt their career (especially women) (Hirsh et al., 1990);
- It is dependant on the availability of key people for the purpose and their willingness to release or share their talent (Leibman et al., 1996);
- It is a lengthy and time-consuming process. Byham (2002) notes that succession management can consume a significant number of executive hours each year.

Although elements of knowledge sharing and transfer are identified from the review, the main focus of succession planning and management is still the transfer of general management skills and to prepare a person to take up a particular position. Therefore, it is not surprising that a range of problems are identified when succession planning and management are assessed against the capability to support 'live' capture and reuse of knowledge.

3.3.8 Reassignment of people

This method is based on the assumption that the knowledge acquired from one project can be transferred by reassigning the people involved to another project (Kamara *et al.*, 2003).

Facilitation of knowledge capture and reuse

Reassignment of experienced staff or experts to other projects can create the opportunity for the transfer of more tacit knowledge through the face-to-face or people-to-people interactions among the members of staff. Less experienced members of staff in particular may be assisted and supervised in carrying out their work by the experienced staff reassigned to the project where knowledge can be captured through observation and mirroring the experienced staff.

Issues and capability to support 'live' capture and reuse of project knowledge

Reassignment of staff inherits some of the shortcomings of the face-to-face interactions. These include:

- Vulnerability to staff turnover;
- The transfer of knowledge is likely restricted to a smaller group of people who have the opportunity to interact with the knowledge provider;
- The willingness of the knowledge providers to share knowledge.

These suggest that it also may not be a suitable tool to facilitate the 'live' capture and reuse of project knowledge.

3.3.9 Knowledge bases

Knowledge bases are repositories that store knowledge about a topic in a concise and organised manner (Ruikar *et al.*, 2003), such as lessons learned and best practices. Knowledge bases are distinguished from the knowledge bases of expert systems which incorporate rules as part of the inference engine that searches the knowledge bases to make decisions (Ruikar *et al.*, 2003).

Facilitation of knowledge capture and reuse

Knowledge in the knowledge base can be captured through very formal sessions specifically conducted for the purpose and from voluntarily contributions from members of staff. An example is found for the former approach: A two-hour session of 25 people worldwide was conducted in an organisation to capture lessons learned, with facilitators and some people specifically assigned to codify knowledge (Leavitt, 2003). The knowledge was documented in various forms ranging from a Gartner-like report to a magazine article. These were subsequently validated by participants and made available to all employees through the company's knowledge base (Leavitt, 2003). The details captured may include where the idea originated, a brief description of the practice, the savings it achieved and the name and phone number of a contact from whom more information can be obtained (Dixon, 2000: p. 56).

For the voluntary approach, it is implemented by encouraging members of staff working on a project to capture and contribute lessons learned from a project into a knowledge base in a predetermined format from time to time (Eppler and Sukowski, 2000). The entries are flagged according to their possible impact to the team's success, that is high, medium or low. To facilitate the use of the knowledge base, simple electronic forms (e.g. best practice templates) and aids (search engines) are normally offered (Heisig and Vorbeck, 2001). This method helps to build up the team's collective memory (or knowledge) which can be consulted (or reused) before critical events (Eppler and Sukowski, 2000) and also in other projects.

Major organisations that have already employed such methods are NASA, the US Army, Siemens Information and Communication Networks, British Aerospace plc, Ford, Texas Instruments, etc. (Eppler and Sukowski, 2000; Dixon, 2000; Heisig and Vorbeck, 2001).

Issues and capability to support 'live' capture and reuse of project knowledge

Nowadays, knowledge bases are often Web-based, or accessible through the Internet and intranet. These types of knowledge bases are capable of facilitating 'live' capture and reuse of project knowledge as they allow people to enter and access knowledge whenever the connection to the knowledge base is available. Current literature does however reveal that although the more explicit knowledge can be captured and shared through knowledge bases, the more tacit dimension of the knowledge still depends mainly on human to human interactions for its sharing and transfer (Heisig and Vorbeck, 2001).

3.3.10   Intranets

An intranet is a company-wide information distribution system that uses Internet tools and technology (Tyndale, 2002). Intranets use World Wide

Web servers and browsers in association with other information retrieval software to deliver information and knowledge to a closed group of users over an organisation's network (ITCBP, 2003).

Facilitation of knowledge capture and reuse

Company procedures, templates, standard statements, frequently asked questions, glossaries and knowledge can be stored in an intranet to preserve organisational memory and for future reuse (Tyndale, 2002; ITCBP, 2003). Intranets allow members of an organisation to access the information or knowledge available from a remote office, a business partner's office and home (ITCBP, 2003).

Intranets can be used to publish (e.g. home pages, newsletters and documents), to search (for a variety of information), to transact (with functionality on intranet pages and other organisational computer-based information systems), to interact (e.g. via discussion groups and other collaborative applications) and to record (e.g. best practices) (Newell *et al.*, 2000). A well-managed intranet can improve cross-organisational communication and enable greater collaboration between different functions (ITCBP, 2003), and hence better sharing of knowledge.

Issues and capability to support 'live' capture and reuse of project knowledge

The intranet's main roles are to provide the necessary ICT backbone (e.g. the network) to facilitate communication across different operating systems and equipment (Newell *et al.*, 2000), and Web-based applications. It mainly depends on other software applications (e.g. knowledge base and groupware) which run on it to facilitate the capture, sharing and reuse of knowledge. Intranets are therefore an enabling technology to facilitate the 'live' capture and reuse of project knowledge. Thus they are not a solution in themselves but need to be complemented by other KM software applications running on the network.

3.3.11 Groupware

The term 'Groupware' refers to ICTs that supports collaboration, communication, coordination of activities and knowledge sharing amongst geographically dispersed groups of people (Dennis *et al.*, 1996; Robertson *et al.*, 2001). It includes the ability to send and receive email, conferencing, shared scheduling of appointments, workflow management and multimedia document management (Rezgui, 2001).

Facilitation of knowledge capture and reuse

Groupware may support a single workgroup on a single LAN, or it may support a number of workgroups and LANs together (Duffy, 1996).

Groupware allows the actors collaborating on specific tasks to exchange ideas, helps to keep track of the project memory and record all its learned lessons in a way that promotes reuse (Rezgui, 2001). It improves information flow to enhance organisation learning and creativity (Bhatt *et al.*, 2005).

Groupware can be useful for the exchange, coordination and articulation of low-level information and explicit knowledge, particularly if the project members are geographically dispersed (Robertson *et al.*, 2001). In addition, recent developments have allowed data mining within groupware's databases to identify potentially valuable knowledge patterns (Bhatt *et al.*, 2005). Furthermore, to an extent, it may also help to capture, store, retrieve and distribute part of tacit knowledge in the form of rituals, histories and organisational stories (Bhatt *et al.*, 2005).

Issues and capability to support 'live' capture and reuse of project knowledge

Robertson *et al.* (2001) argue that groupware is less efficient for the communication and exchange of more complex tacit knowledge. While a groupware's database provides relevant knowledge in a timely fashion, often clarification may still be required on particular complex information through face-to-face or telephone conversation (Robertson *et al.*, 2001). Furthermore, in many cases the basic functionalities and mechanisms of groupware systems are not sufficient to support users in finding the required knowledge (Smolnik and Erdmann, 2003).

3.3.12 Project extranets

A project extranet is a network linking the various parties to a construction project for the exchange and storage of project information in digital form (Hamilton, 2005). Its access is only extended to a privileged user group from those parties or organisations (Watson, 1999).

Facilitation of knowledge capture and reuse

According to Howard (2004) project extranets can help organisations to:

- Share up-to-date documents, files or images with suppliers, partners or customers in disparate locations;
- Work collaboratively by making documents or digital assets available for editing, reviewing, updating, versioning and storing;
- Manage projects in a centralised workspace and track the history of work;
- Provide current versions of frequently updated documents, such as product specifications, inventory summary and design documents.

Project extranets also allow the project data (including the documents uploaded) to be stored permanently for future access (Hamilton,

2005). Ruikar *et al.*'s (2005) research findings reveal that this function can be very valuable to end-user companies as they can be used to resolve future issues of similar nature (e.g. lessons learned from previous projects or stages of the same project can be applied to latter projects/stages). However, Ruikar *et al.* (2005) also state that none of the companies they studied have taken measures to benefit from this.

Issues and capability to support 'live' capture and reuse of project knowledge

Rezgui (2001) states that most project extranets only provide support for document storage, retrieval, versioning and approval and do not handle the semantics of the information being processed in the documents. Therefore, they are less efficient in facilitating the reuse of the knowledge and lessons stored within these documents (Rezgui, 2001). This may account for the lack of effort from end-user companies to benefit from reusing the knowledge captured in the documents, as revealed by Ruikar *et al.* (2005). Furthermore, there are several issues that undermine the use of project extranets. These include the security of the information stored, necessary culture change for adopting the technology, cost of implementation and legal issues as to the ownership of data (Ruikar *et al.*, 2005).

3.3.13   *Case-based reasoning*

Case-based reasoning (CBR) is a problem-solving approach that relies on past similar cases to find solutions to problems (Kolodner, 1993). According to Kolodner (1993), a case is a 'contextualised piece of knowledge representing an experience that teaches a lesson fundamental to achieving the goals of the reasoner'. In most CBR systems, the internal structure can be divided into two major parts: the case retriever and the case reasoner (Shiu and Pal, 2004). The case retriever's task is to find the appropriate cases in the case base, while the case reasoner uses the retrieved cases to find a solution to the given problem description (Shiu and Pal, 2004). According to Aamodt and Plaza (1994), CBR essentially consists of four processes (i.e. the four REs):

- *Retrieve* the most similar case or cases;
- *Reuse* the information and knowledge in that case to solve the problem;
- *Revise* the proposed solution;
- *Retain* the parts of this experience likely to be useful for future problem solving.

Although not included in the four REs, Aamodt and Plaza (1994) acknowledge that the representation of cases is crucial to make the case search and matching processes of the case retriever and case reasoner both effective and reasonably time efficient. The representation of cases covers what to store in a case, the finding of an appropriate structure for describing case contents and the decision on how the case memory should be organised and indexed for effective retrieval and reuse (Aamodt and Plaza, 1994).

Facilitation of knowledge capture and reuse

CBR has been used with positive results for customer service and help desk applications (Belecheanu *et al.*, 2003). In terms of knowledge capture and reuse, the roles of CBR are as follows:

- It reduces the knowledge acquisition task by eliminating the need to extract a model or a set of rules, as required in model-/rule-based systems (Shiu and Pal, 2004). CBR's knowledge acquisition tasks involve mainly the collection, representation and storage of existing cases (Shiu and Pal, 2004).
- It provides flexibility in knowledge modelling (Shiu and Pal, 2004). CBR systems deal with case-specific knowledge and do not require that the domain be modelled in rules (Belecheanu *et al.*, 2003).
- It may help to discover and retrieve quality design solutions that are stored in a specifically designed knowledge base (Cirovic and Cekic, 2002).
- It helps to avoid repeating all the steps that need to be taken to arrive at a solution (Shiu and Pal, 2004). The ability of CBR to help in modifying a previous solution to solve a new problem, instead of creating a solution from scratch, leads to significant time savings (Shiu and Pal, 2004).

Issues and capability to support 'live' capture and reuse of project knowledge

Belecheanu *et al.* (2003) note that CBR systems can be costly to develop and implement, demand substantial technical training and support and their efficiency depends on the willingness of people to use and improve the system on a daily basis. Furthermore, the applicability of the solutions retrieved cannot be guaranteed when the problem is too complex and covers a wide scope (Belecheanu *et al.*, 2003).

3.3.14 *Text mining*

Text mining, also known as text data mining or knowledge discovery from textual databases, refers to the process of extracting interesting and non-trivial patterns or knowledge from text documents (Tan, 1999). The difference between text mining and data mining is that in the former the patterns are extracted from natural language text rather than from data sets (Hearst, 2003). Text mining links together the extracted information to form new facts or new hypotheses to be explored by further research (Hearst, 2003).

Facilitation of knowledge capture and reuse

Tan (1999) notes that text mining tools can help in:

- Organising documents based on their similarities and presenting the groups or clusters of the documents in certain graphical representation. Tan (1999) also points out that some tools can map the links between concepts in the document collection;
- Analysing texts. This covers extraction, retrieval, categorisation and summarisation of texts and information (Tan, 1999). Some of the tools can 'learn' the relationships between words and phrases automatically from sample documents and guide the users to construct searches (Tan, 1999).

Hearst (1999) adds that text mining may aid the discovery of unknown information or the finding of answers to questions for which the answer is not currently known. Text mining may involve finding the unexpected patterns and trends among text articles or information (Hearst, 1999).

Issues and capability to support 'live' capture and reuse of project knowledge

Text mining is mainly concerned with searching for and identifying unexpected trends, and unknown information and knowledge. Its main purpose is the support of the knowledge discovery process in large document collections (Karanikas and Theodoulidis, 2002), where knowledge captured in other formats (e.g. drawings and video clips) are ignored. Furthermore, it has been suggested that text mining technology is still undermined by the inability of computers to understand text as humans do and in sorting out ambiguous words (Leong et al., 2004).

3.4   Soft issues in KM

IT has been the centre of many KM initiatives (Walsham, 2001) probably because of the growth in knowledge-based expert systems in the eighties and early nineties (Kamara et al., 2003). However, it is now acknowledged that IT tools alone do not stimulate individual affection for the generation of knowledge (Neve, 2003), and leveraging knowledge exclusively through ICT is often hard to achieve (De Long, 1997; Walsham, 2001). The impacts of 'soft' or non-IT issues on KM are discussed below. The findings are grouped into people, organisational and cultural issues.

3.4.1   People issues

People play a vital role in knowledge capture and reuse practices as only people are regarded to be able to learn in organisational learning theory, and ultimately organisations learn from people (Hwang, 2003). The various people issues identified are described as follows:

Willingness to share knowledge

Ardichvili *et al.* (2003) have identified that people's willingness to share knowledge is influenced by their perception as to the ownership of knowledge; that is either viewing it as a public good or a personal belonging. When knowledge is regarded as public good, knowledge exchange is motivated by moral obligation and community interest and not by a narrow self-interest (Ardichvili *et al.*, 2003). On the other hand, if knowledge is regarded as a personal belonging then knowledge hoarding or reluctance to share knowledge is envisaged.

Self-confidence

The lack of self-confidence and the confidence in the knowledge to be shared by employees have also undermined knowledge capture and sharing practices. Ardichvili *et al.*'s (2003) findings reveal that employees may hesitate to contribute their knowledge out of fear of criticism or misleading others, being unsure that their contribution is important, or completely accurate, or relevant. Furthermore, they may simply think that they have not earned the right to post and share their knowledge within the organisation.

Trust

The lack of trust has been identified by Mason and Pauleen (2003) as one of the barriers to implementing KM. This finding corresponds to Ardichvili *et al.*'s (2003) contention that people are less reluctant to share knowledge if they think that others will not misuse their knowledge (e.g. taking undue advantage of confidential information and using the posted information to personally attack those who posted it). In addition, trust is also in turn built upon people's confidence that the knowledge shared is reliable and objective.

Shared meaning

The transfer of knowledge demands the existence of shared meaning, which is a shared mental model or system of meaning that enables others to understand and accept the knowledge, and apply another's insight to their own context (Bresnen *et al.*, 2003). The time required for developing the shared meaning for inter-project knowledge transfer is a main issue of concern as project teams are normally temporary and culturally differentiated (Bresnen *et al.*, 2003).

Personal KM capability

Personal KM capability refers to the capability of employees to capture and share knowledge, and use the IT-based KM system for the purpose. This capability can greatly influence the efficiency of knowledge capture and reuse practices, particularly when IT-based KM systems are used. However, it is argued that personal KM capability is something which people can improve over time, assisted by appropriate training.

Staff mobility and turnover

Change in the membership of a project team during the course of a project and high staff turnover often result in organisational knowledge fragmentation and loss of organisational learning (Kasvi *et al.*, 2003). Reducing frequent change in project team membership, and the retention of members of staff is therefore crucial in preventing such knowledge gaps from developing (Harman and Brelade, 2000).

3.4.2 Organisational issues

Both inter- and intra-organisational knowledge sharing are affected by issues such as commercial sensitivity of the knowledge, existence of rewards to encourage knowledge sharing and other management issues. The details of the main issues are as follows:

Commercial sensitivity and security of knowledge

Commercial sensitivity and security of knowledge is critical for inter-organisational knowledge sharing (Barson *et al.*, 2000). The corporate security restrictions imposed on the posting of information/knowledge (Ardichvili *et al.*, 2003) may indirectly discourage people from sharing their knowledge where the boundary of such restrictions is not made clear.

Creation of a reward and incentive structure

Current literature suggests that the impacts of incentives on knowledge capture and reuse are two-sided. From one perspective, incentives and rewards have been identified as the key success factor in the knowledge sharing process (Hansen *et al.*, 1999; Eppler and Sukowski, 2000; Hall *et al.*, 2000; Robinson *et al.*, 2001), in sustaining a knowledge sharing culture (O'Dell *et al.*, 1998; Neve, 2003) and in encouraging people to engage in knowledge-based roles, activities and processes (Zack, 1999). From another perspective, an ill-designed system and/or the lack of an incentive structure can also act as disincentives and lead to knowledge hoarding

(Hall *et al.*, 2000). The dominating outcome-oriented approach where one's reward is judged from the amount of work done on the codification and dissemination of knowledge (Hall *et al.*, 2000) is very individual-based and does not encourage collaborative behaviour (Walsham, 2001). Therefore, the more balanced 'scorecard' incentive system suggested by Goh (2002) whereby both collaboration with other teams and sharing of knowledge are taken into account to avoid internal competition is likely to be a better option.

Allocation of resources

Hall *et al.* (2000) argue that conflicts exist as the strategic benefits from knowledge transfer are accrued by the organisation as a whole, but the tactical cost for the resources (budget and staff time) associated with knowledge capture is borne by the individual projects. Consequently, knowledge capture and sharing may be sacrificed particularly when the risk of project cost overrun is greater than the risks associated with the loss of organisational knowledge (Hall *et al.*, 2000). It is therefore suggested that the cost incurred is treated as overhead expenses for the organisation to solve this problem (Hall *et al.*, 2000).

Company policy towards lessons learned

Knowledge sharing should both cover the best practices learned and the mistakes made (i.e. lessons learned). However, people are reluctant to admit mistakes (De Long, 1997) and the organisation's disciplinary procedures further discourage the sharing of this kind of knowledge (Hall *et al.*, 2000). Heisig *et al.* (2001) suggest that organisations should recognise that errors are an essential factor in the process of learning and should hence maintain a 'culture of errors' where making errors are tolerated to an extent to encourage the sharing of lessons learned.

3.4.3   Cultural issues

De Long (1997) contends that culture has a major impact on the implementation of any KM strategy, and it comprises the following three elements:

- *Values* that indicate what an organisation's members believe is worth doing or having. They indicate preferences for specific outcomes or behaviours, or what the organisation aspires to achieve.
- *Norms* which are the shared beliefs about how people in the organisation should behave, or what they should do to accomplish their work. It represents the expected patterns of behaviour. For example, they

describe how employees actually create, share and use knowledge in their work.

- *Practices* which are the formal or informal routines used in the organisation to accomplish work.

Organisational cultures can either be supportive or negative towards KM as follows:

Supportive and negative knowledge cultures

A supportive KM culture where employees are motivated to share their knowledge and use external knowledge for their own activities is essential (Walsham, 2001). Culture has even influenced the performance of IT-based KM systems. As De Long (1997) asserts, IT-based KM systems will be implemented and used effectively only to the degree that a culture is aligned to support the objectives for KM. However, such a KM supportive culture may not readily exist within an organisation, and the creation of one very often requires the alteration of existing organisational culture (De Long, 1997).

Negative cultures which include resistance to search, receive and share knowledge are found to have undermined organisations' KM practices (Ladd and Heminger, 2002). In the context of the construction organisation, negative culture such as employee resistance may inhibit knowledge sharing as people feel insecure about their job situation and do not trust their employers (Robinson *et al.*, 2001).

Cross-cultural knowledge sharing

Cross-cultural knowledge sharing may cover the sharing of knowledge across organisations, societies and countries. Knowledge sharing across organisations is complicated and often unsuccessful due to the differences in educational background, skills base and approach to the coordination of work (Walsham, 2001). This is critical in the context of the construction industry as construction projects normally involve various organisations with different expertise, objectives and working cultures. Bresman *et al.* (1999) contend that the problems associated with knowledge sharing will increase with geographical and cultural distance. Therefore, knowledge sharing across organisations will be even more challenging if it involves organisations from different societies or countries. This is substantiated by Hutchings and Michailova's (2004) research findings which reveal that very often successful knowledge sharing across organisations from different countries entails a detailed study of the relevant countries' cultures in advance and involves extensive relationship or network building in order to facilitate it.

3.5   Summary

A range of reusable project knowledge items and learning situations have been identified from case studies and the literature. A review of the various KM technologies and techniques suggests that a knowledge base is a potential KM tool capable of facilitating a methodology for 'live' capture and reuse of project knowledge in construction. The various soft issues that may influence the implementation of a KM system within an organisation are also identified. All these are explored further in the Chapter 4.

4 Collaborative Learning in Construction

This chapter provides a general review of collaborative learning (CL) and then explores how it can be related to and harnessed by the construction industry. It proposes a CL strategy that involves the capture and reuse of knowledge in projects that will reflect both the organisational and human dimensions to deliver a better result. The key considerations for a heterogeneous project team are also discussed.

4.1 Introduction

The concept of collaborative learning (CL) can be traced back to the nineteenth century during the scientific revolution when there was a gigantic Western exercise in learning (Lessem, 1990). However, it was not until the early 1970s that researchers adopted organisational learning as a way of competitive advantage (Huber, 1991; Gill, 1995). The late 1990s saw CL become increasingly important as a result of its association with the concept of knowledge management (KM) (Ruggles, 1997; Jackson, 1998; Patel *et al.*, 2000; Tiwana, 2000; Tsui, 2002; Nonaka and Toyama, 2003; Cuthell, 2005; Hernández-leo *et al.*, 2006).

This move towards more collaborative working within the industry is a welcome one, given the numerous problems that have resulted from the industry's fragmentation and adversarial nature (Anumba and Evbuomwan, 1998). CL is increasingly being seen as an appropriate vehicle for implementing the changes required to integrate both construction project participants to improve the overall project delivery and help organisations in their future projects. Each expert understands the project slightly differently; they may react differently to what could be the same situation. This highlights the importance of getting people together to establish a shared understanding of any problem situation and the potential pathways for action. When people feel that they have had the opportunity to participate in planning the project, they are likely to buy into the changes that may be required of them. In construction, creating a learning approach has become the goal of many organisations (Patel *et al.*, 2000). Contemporary business challenges demand a new kind of learning, one that goes beyond problem solving within an organisation and instead focuses on imagining possibilities and new ways of looking at entire businesses globally (Allen *et al.*, 2001) and also looking at the relationships between intra- and inter-organisations (Bresnen, 1996) involved in a construction project. The ability for organisations to learn from one another is critical to sustaining and building competitive advantage (Harvey *et al.*, 1998).

4.2   Collaborative learning

Over the past few decades, social science understanding of what motivates changes in human behaviour recognises that people are active sense-makers (Hasan and Gould, 2001; Thomas *et al.*, 2001), who are continually assessing their environment and acting according to their interpretations of the situation. Because each individual or group experiences the world slightly differently, they may react differently to what may be the same situation. This highlights the importance of getting people together to establish a shared understanding of any problem situation and the potential pathways for action (Holt *et al.*, 2000). When people feel that they have had the opportunity to participate in planning future change, they are likely to buy into the changes that may be required of them.

The continuously changing collaborative environment necessitates a well-delineated orientation to bring about learning. Learning is a social and interpretive activity in which multiple members collaboratively construct explanations and understandings of materials, artefacts and phenomena within their environment (Jones, 1995). It is the result of active engagement in and with the world coupled with reflections upon the relationship between ideas, actions and outcomes. As such, learning-as-interpretation is a deeply embedded active and reactive process. Collaborative activity presents an opportunity for reflection and interpretation of events by providing a shared context for the interpretation of individual experience. Interpretations evolve around artefacts and narratives (Harvey *et al.*, 1998), and experiences take on meaning within communities of practice (Wenger, 1998). Learning also involves developing new understanding. Research in the area of cognitive and behavioural sciences at the individual level describes the learning process as involving the acquisition and interpretation of knowledge (Fiol, 1994). The process need not be conscious or intentional, nor need it necessarily increase the learner's effectiveness or visibly change the learner's behaviour (Pentland, 1995). Rather, learning is the process of modifying one's 'cognitive maps or understandings', thereby changing the range of one's potential behaviours (Fiol, 1994). Thomas *et al.* (2001) argues that the ability of organisations to learn is directly linked to the way the environment and information are interpreted and addressed by the organisation.

It is useful to see learning as being made up of two components – its process and the outcomes of that process. Change can be observed as an outcome of learning. This, in turn, must be viewed as an accumulative process which builds on existing practices and norms through interactive learning. While information is central to this process, learning also needs to be supported by other conditions. Key among these is the need to build and maintain trust among different parties involved. Other processes will also be required to manage forums which enable the development of a shared understanding such that stakeholders can quickly and effectively place problems and information in their wider context (Allen *et al.*, 2001).

The topic of collaboration has received attention ever since, yet conceptually, CL is not particularly well understood. In providing a topology of the CL concept, this book supports Harvey *et al.* (1998)'s contention that 'Collaborative Learning' originated from organisational learning. They argue that the work on organisational learning has gone in and out of style, since the early work of Argyris and Schon (1978), whose work on organisational learning was credited as the original birth of Learning in business management. In another precedent, many labels have been found referring to CL as a classroom technique, where people come together in groups (Holt *et al.*, 1995; Panitz, 1996; Kaplan, 2002). Many of these proponents working in this area have suggested that CL is a way of dealing with people which respects and highlights individual group members' abilities and contributions. Most argue that there is sharing of authority and acceptance of responsibility among group members for the group actions. They also all agreed that the underlying premise of CL is based upon consensus building through cooperation by group members, in contrast to competition that exist in a normal setting. However, the general consensus from both schools of thoughts is that practitioners can apply this philosophy in the classroom, organisations, project undertaken, at committee meetings, with community groups, within their families and generally as a way of living with and dealing with other people.

Digenti (1999) in defining CL, argued that it is a business practice that is aimed at discovering explicit and tacit collaboration tools, processes and knowledge, experimenting with them and creating new knowledge from them. By CL, the authors argue that it is an interaction of two or more people that engage in value-creating and knowledge capture activities based on improving, practicing and transferring collaboration skills and expertise both within the group and to the organisation or group of organisations to which they belong (Udeaja *et al.*, 2005). CL employs experimentation, methods and approaches that emerges from the present situation and evolve as they are practiced. This method of exploration has been termed action research, originating from the work of Lewin (1946). Organisational learning and culture concepts underpin the functioning of CL. These methods allow organisations to move across boundaries fluidly and to ensure that the learning that takes place in one group is transferred back to the organisation. The practice of these approaches demands a fundamental shift of tapping into expertise real-time and creating an environment that allows this to happen (Digenti, 1999). Digenti (1999) further argued that the shared goals of this learning group is the generalisation of the information acquired for the benefit of individuals involved.

However, the increasing use of collaborative approaches to research and development pose new challenges for decision-makers and evaluators. Because these programmes are designed to be responsive to changing community needs and social goals, one of the most pressing challenges is to develop participatory and system-based evaluative processes to allow for ongoing learning, correction and adjustment by all parties concerned. In particular, because they are often focussed towards the development of change, which takes time, rather than shorter term and more easily defined outputs they are often hard to measure (Allen *et al.*, 2001). The next section of this book will discuss the problem issues that CL can address and look at the previous research in CL from a general point of view of the two underlying philosophies, to show how various researchers in this area have conceptualised 'Collaborative Learning'.

4.3   CL in construction

Latham (1994) and Egan (1998) identified that the project nature of the construction industry pose great challenges and barriers to learning. Many practitioners and researchers in the industry have acknowledged the limitations of current approaches to managing information and knowledge related to and arising from a construction project (Fruchter *et al.*, 2000; Rezgui, 2001; Lima *et al.*, 2002). These limitations are due to several technical-, human- and business-related factors. However, these outline the requirements and problems of learning in construction projects. A detailed outline of these problems is documented in CAPRIKON Report (2004). The key issues can be summarised as follows:

- Construction projects are temporary multidisciplinary organisations where the scope for continuous interaction among project participants, after the end of a project is limited.
- Much of construction knowledge still resides in the heads of individuals, or at best, exists in an informal and unstructured form that makes it difficult to comprehend and exploit.
- Different discipline solutions interact with each other. The process of identifying shared interests is ad hoc and based on participant's imperfect memories. This error-prone and time-consuming process rapidly leads to inconsistencies and conflicts.
- Experience gained while solving a problem during the course of project is not adequately transferred to other people or incorporated as the project progressed. Team members complete the task and take any learning along with them to new teams. Partial loss of project memory takes place if team members are not going to use the knowledge and information that they acquired from previous project again on a new project.

- The end of the project marks the end of the learning of whole team. Post-project evaluation is usually conducted by participating organisations to a project, and is useful in consolidating the learning of people involved in the project under review.
- There is a common problem of insufficient time for post-project evaluation to be conducted effectively (if conducted at all), as relevant personnel would have been moved to other projects. Furthermore, it does not allow the current project to be improved by incorporating the lessons being learnt as the project progresses.
- There is also the problem of loss of important information or insights due to the time lapse in capturing the learning. Moreover, in consolidating the learning of people involved, post-project evaluation is not a very effective mechanism for the transfer of knowledge to non-project participants. It is also limited in scope, in that the perspective is that of members within only one of the participating organisations to the project.
- The use of long-standing (framework) agreements (e.g. within a partnering contract) with suppliers to maintain continuity in the delivery of projects for a specific client is also designed to ensure that the learning by individuals and firms is reused on future projects. However the reliance on people (even within a framework agreement) makes organisations vulnerable when there is a high staff turnover.
- The use of framework agreements also cannot guarantee that the learning of individual firms participating in the agreement is shared to other participants (for the benefit of the project), since these firms can be in competition elsewhere (e.g. on other projects) and may not want to divulge 'secrets' that might weaken their competitive advantage.

These factors have not merely inhibited effective KM; but they have inhibited the industry's ability to capture, learn and reuse project knowledge for improved performance. If learning across projects takes place, it will ensure that experiences are accessible through informal networks. Also as problems happen, solutions can be devised, effectively capturing problems, causes and how these are carried out. This could also ensure that proper project learning occurs on projects and that documentation-based methods are adopted to capture project knowledge and information as it happens. Improvements in project procurement using CL approach can reduce the construction period and help clients save cost. Some of these improvements can be accomplished through better capture and reuse of learning during the project life cycle. From the above problem definition, the requirement is to use appropriate knowledge and information infrastructure, and improve collaborative working between members of a project team. The next section will discuss the previous research in CL from a general perspective. This will show how researchers have conceptualised CL from their own field.

4.4   Previous research in CL

The first great industrial revolution in the nineteenth century and the scientific revolution that preceded it, was the first exercise in learning. However, it was not until the latter part of the twentieth century that the idea of a 'learning organisation' gained prominence. The impetus for this has in fact come from the idea of capturing learning and disseminating the learning throughout the organisation by Argyris and Schon, 1978. Many studies have followed in this vein, dealing specifically with organisational learning (Huber, 1991; Pentland, 1995; Harvey *et al.*, 1998; Holt *et al.*, 2000; Patel *et al.*, 2000; Bhatt and Zaveri, 2002; Strati, 2007; Spender, 2008), project-based learning (Björkegren, 1999; Thomas, 2000; Bresnen *et al.*, 2002; Fong, 2003; Kamara *et al.*, 2005) and CL (Gokhale, 1995; Holt *et al.*, 1995; Panitz, 1996; Sadler-smith *et al.*, 2000; Allen *et al.*, 2001; Kaplan, 2002; Udeaja *et al.*, 2005; Hernández-leo *et al.*, 2006). There are several ongoing research projects investigating aspects of learning between members of the collaborating organisations. The research projects cut across various disciplines and sections including construction, manufacturing, education and other engineering areas and employ variety of strategies and concepts. The remainder of this section will look at the previous works from a general point of view of the underlying philosophies, to show how the concept of 'Collaborative Learning' has being developed.

Harvey *et al.* (1998) work draw from previous work in this area, where organisational learning is based upon an environment which encourages the development of processes for acquiring knowledge, disseminating and interpreting information, as well as creating an organisation memory. They agreed that to effectively simulate organisational learning, management must create an infrastructure to encourage, support and document the learning taking place. Based on this, they made the case that a learning organisation requires a highly flexible information infrastructure that allows individuals to pull information out of existing repositories as needed. In all they presented a phased model of intra-organisational learning, which is prescriptive in nature. It articulates the communication flows and information sharing which should take place across organisational structures to enable the existing culture to evolve into an ongoing learning environment. The outcome of their implementation is a four-phased learning culture that is initiated from top-level executives and passed down to the functional management level at phase one. The second phase deals with expansion of learning environment across functions thereby inculcating a learning culture throughout the organisation by promoting cross-function communications, multifunction team problem solving, and eventually, cross-functional learning. The third, encourages learning between divisions within the same organisation, this basically address the political barriers to cooperation that are frequently encountered when spanning divisional boundaries. Finally, the fourth phase bolster learning between organisations owned by a single company – learning in this phase takes place between two distinct organisational entities that share common corporate ownership.

Ruhleder and Twidale (2000) presented reflective CL in the Web that draws from a master class. The work explores a connection between an established form of intensive, face-to-face teaching, the master class employed in the Arts, and new possibilities for organising online learning experiences for a distributed class, including classes teaching aspects of science or engineering. They argue that the format and pedagogical goals of a master class embody a set of principles compatible with Dewey's conceptualisation of reflective learning. These classes take place within a community of practice that supports ongoing CL that bridges the boundaries of the classroom. The work draws on a choral conducting master class as an illustration, and then uses this as a springboard to illustrate how these principles are being brought into virtual settings, using as an example an online distributed class as interface design. They further presented how the virtual degree program to which this distributed class belongs supports certain educational experiences. They concluded by arguing that their technique has a broader implications and opportunities for creating robust online venues for CL. Thus they are taking the organisation of the traditional music master class as a means of addressing a pedagogical need for teaching interface design at a distance, rather than face-to-face in a laboratory context, as it would conventionally be taught.

Allen *et al.* (2001) presented Integrated System for Knowledge Management (ISKM) approach that illustrates how such learning-based approaches can be used to help communities develop, apply and refine technical information within a larger context of shared understanding. They used a case study involving pest management to illustrate how CL approaches can be used along with more traditional linear forms of information transfer to support improved environmental decision making. The work discussed the social context and challenges facing those involved in the case study, and then identified possible solutions that can be used to promote a more active form of information management. It also discussed how the ISKM approach can help implement such an active or learning-based approach. The work showed huge potential for using the Internet to support and disseminate experience gained through ongoing adaptive management processes. However, the work conclude that collaboratively developing new management options and strategies through the ISKM process provides interested parties with the opportunity to learn from local experiences gained within enterprise and catchment-level systems. This provide those involved with an appreciation of management concerns and issues, and gives scientists and policy-makers a better feeling of how their contributions fit into the total system. They argued that this holistic approach is important because much of the conflict surrounding many resource management issues arises from different interest groups failing to appreciate the perspectives and values inherent in the actions of others. If these groups can be encouraged to share their experiences and viewpoints, there will be a greater understanding of why these differences exist.

Thomas *et al.* (2001) presented strategic learning aimed at generating learning in support of future strategic initiatives that will, in turn, foster knowledge asymmetries that can lead to differences in organisational performance. Their argument is that creating and disseminating knowledge for strategic purposes within and across level of analysis appears as a recurring theme in most literatures. Based on this, they claimed that the primary motivation for their research is to identify illustrative organisational practices and processes that contribute to performance-enhancing strategic learning. Their second motivation was derived from the observation that strategic learning has been conceived of, alternating, as a process to foster continuous radical innovation over the long term, and the focused exploration of anticipated future events and activities. Their review suggested that literatures in this area compliments both perspectives, directly and indirectly, suggesting that future inquiry into strategic learning must also include investigation of the roles of sense making, KM and information transfer processes. In this sense, the understanding of interpretive processes, subsequent learning and transfer of lessons learned need to be combined to enable strategic learning. Such understanding is critical to optimise allocation of organisational resources in a strategic and innovative learning environment.

Under this conditions, they chose an investigative technique referred to as theoretical sampling, wherein a case is selected as a unique example of a particular phenomenon to bring key dimensions to light. They selected an appropriate case study (CALL) as their context for inquiry into strategic learning. Based on their analysis, four characteristics of 'Strategic Learning' became apparent: Data collection efforts are targeted; it is timed to coincide with the strategic action horizon of the firm; it leverages the organisation's ability to generate, store and transport rich de-embedded knowledge across multiple levels for the purpose of enhancing firm performance and it has institutionally based sense-making mechanisms in place with associated well-defined validation processes. These characteristics was used to craft a set of propositions to guide future inquiry, and to build a theoretical model based on those propositions, which frames how strategic learning can be manifested. The findings of this research provide a rich theoretical description of how one organisation is developing the systemic capability to rapid learning from ongoing practice and to create foreshadowed knowledge of future events. In doing so, it stands at one end of several dimensions that researchers can use to understand strategic learning in other organisations, and that practitioners can use as design parameters to build variants of this system.

Hamada and Scott (2001) discussed a CL model for distance learning courses. They claim that learning is extricable intertwined with multi-directional activities such as work and play, and that learning is essentially a social activity. They argue that position of learning is a process of applying it, because knowledge is temporary, developmental and socially and culturally mediated. On these bases, they developed a collaborative learning and teaching (COLT) model to engage students. The model required the students to conduct their own research for knowledge creation at local sites; connects students with different cultural backgrounds for direct cross-cultural interface; the collaboration for co-knowing is international; the students employ a much wider range of communication tools; the final project is not only edited collaboratively, but also presented collaboratively and it intentionally creates a learning environment where participants need to manage uncertainty and uncertain knowledge. In all, the COLT model allows collaborative groups to execute tasks that are too complex for one individual to undertake. It provides opportunity for students to participate in cross-cultural group dynamics, to articulate, explicate and defend their ideas and hidden motives, and to manage their work flow amid high degree of uncertainty about how the project should be done. At the end, they must create an intellectual product collaboratively.

Organisational learning as described by Patel *et al.* (2000) is the ability of the organisation to collect and use information so that members exploit it to learn and to improve performance. They went on to say that learning is something that pervades every individual's life in one form or another. Organisations may be capable of learning and such organisational learning may in turn impact upon various aspects of an organisation's performance (Patel *et al.*, 2000). The research discussed the role of IT in capturing and managing knowledge for organisational learning on construction projects – known as KLICON. The KLICON project's aim was to improve the understanding of the role of KM and how it adds value in the built environment. This was achieved by studying the participating industrial organisations to analyse how experience and best practice were being captured. The project also used IDEF0 and information models in EXPRESS to enhance understanding of generic construction knowledge and specific project knowledge. They also evaluated the issue of live project and identified key KM tools. They concluded that KLICON provided an understanding amongst construction practitioners of how knowledge is gained and learning is formalised across the organisational interfaces within a project. The role and appropriateness of IT tools for knowledge capture and management was also clarified.

Sadler-smith *et al.* (2000) in suggesting that learning is one of the keys to sustained competitive advantage developed a model for CL in small firms. They argued that resource constraints within smaller firms may mean that they sometimes fail to maximise the potential of learning within their organisation. The aim of their work is to develop a collaborative model of small firm learning and its implementation in a number of organisations in the South West of the UK, as well as to address a manageable number of high-priority learning issues. The model took as its starting point a diagnosis of learning needs at the individual and organisational levels. They argued that managerial and organisational learning are two areas in which there is scope for potentially valuable work in the smaller firm sector. The model consisted of a number of elements. The first used MCI standard as a framework to provide subjective and objective assessment of the participants' learning needs. The second assessed the learning orientation of the firm by using a company learning profile. The third used the SWOT framework to identify learning needs at the organisational level. The last component consisted of developing a learning programme that was specific to individual and organisational learning needs. The research concluded that the model represents a new approach for facilitating learning in smaller firms. They claimed that a number of objectives have been achieved: a package of diagnostic tools has been successfully developed; self-profile generated value data for the participants that made them to reflect on their own competencies and the capabilities of their organisations and they argued that it also demonstrated its utility as a research tool. The also claimed that the MCI standards were useful as a loose framework. The argued that the SWOT framework also acted as an assessment and learning function. They showed that the feedback from participant firms was that the SWOT provided an effective tool for assisting in identifying organisational and managerial actions that promoted collective learning.

Holmqvist (2003) in describing intra- and inter- organisational learning processes argued that organisational learning literature has so far focused primarily on intra-organisational learning processes. The aim of the study was to explore the way a company learnt both independently and together with its business partners using empirical comparison. He reported some of the findings from a case study relating to the Scandinavian software producer Scandinavian PC systems (SPCS). SPCS was considered to be an interesting object to study from a learning perspective, due to the alleged intensity of its product development and the number of partners. It also met the practical requirement of willingness to participate in the research. The research concentrated on four product development projects both within SPCS and between SPCS itself and between the company and its partners, which involved both the refinement of existing products and the elaboration of entirely new ones. He claimed that product development could help to understand about how both employees within the respective companies interacted with one another and how employees between the companies did so. Product development project allowed for the study of learning both within and between organisations. In studying organisational learning processes, he adopted the traditional approach of organisational learning theory and concentrated on the production and re-production of organisational rules. These included highly formalised and written rules and routines, as well as more tacit and informal conventions, roles and codes based on experiences.

He assumed that learning had taken place when a set of individuals started to behave according to some tacit and explicit rules. He also studied how organisational members from the various companies produced experiential rule through bargaining when they were confronted with specific situations in new or ongoing product development projects. He argued that the bargaining that occurred between different professional groups in SPCS, or between employees of the various companies partaking in SPCS's inter-organisational collaboration, was a crucial importance to an analysis of learning dynamics. In conclusion he claimed that the comparison provided an opportunity to discuss how formal inter-organisational collaborations such as strategic alliances can learn on their own account by producing and re-producing inter-organisational rules. He went on to say that the comparison can also contribute to a discussion of the value and conceptual justification of the increasingly common separation between intra-organisational and inter-organisational learning in the literature. However, the study identified one limitation of this empirical focus that excluded potentially important learning processes, such as those arising from organisational crises, or from the recruitment of new employees.

The proliferation of research projects demonstrates the increasing interest of researchers in both academia and industry in the area of learning in project-based environments. However, the construction industry still has a significant gap to bridge to reach best practice in addressing the issues in this area. Fundamental changes are required to address the issues evolving from the previous research and applications.

4.5   Implementing CL in construction projects

Several researchers have described project development as a knowledge-intensive and learning activity. Project development often involves cross-functional linkages, where different participants join a team with different viewpoints. Such teams are often characterised according to the risk and synergy resulting from their interaction with other team members (Huang and Newell, 2003). This interaction brings in the need to organise, integrate, filter, condense and annotate the collaborative data and other relevant information that these team members contribute (Fong, 2003). Creating new knowledge and learning is fundamental to project development (Huang and Newell, 2003). A project as discussed by Hamilton (2001) can be considered as a package of features and benefits, each of which must be conceived, articulated, designed, constructed and maintained. The development of this constructed facility can be viewed as a new product development, with customers or end-users purchasing or using the facility (Fong, 2003). Fong (2003) argues that the development of a new product entails the application of knowledge and learning to new problem-oriented situations, thus requiring uncertainty reduction.

The same applies in construction, with each project unique in itself in terms of design and construction, and the many constraints, the construction industry faces (due to limited space, increasing project complexity, limited budgets, tight programmes and constant demand for facility innovation). Project teams are also faced with the challenges to utilise diverse knowledge and create new knowledge in order to meet stringent requirements and fulfil ever-changing needs. Project team members have to incorporate new information into their understanding in order to solve the technical challenges they face. Thus, learning and knowledge capture is inherent in the work they do (project development). Kamara *et al.* (2002a) in discussing the CLEVER project identified that among the various initiatives for addressing the challenges facing the construction industry, it is now recognised that the management of project and organisation knowledge is necessary if construction businesses are to remain competitive, and adequately respond to the needs of their client. They went on to say that failure to learn and transfer project knowledge, especially within the context of temporary virtual organisations, will lead to reinventing the wheel, which will amount to wasted activity and impaired project performance.

In a project-based environment, such as construction industry, it is highly desirable that lessons learnt or captured from one project are put into use in the same project or on subsequent projects, achieving reduction in project times and subsequent efficiencies (Udeaja *et al.*, 2005). Kamara *et al.* (2002a) argued that the need for learning is fuelled by the need for innovation, improved business performance and client satisfaction within the dynamic and changing environment. Project-based organisations ought to benefit from the inherently innovative nature of project tasks. Since projects characteristically involve the development of new products and new processes, there are obvious opportunities for novel ideas to emerge and for cross-functional learning to occur, thereby enhancing the organisation's innovative capacity and potential (Ramaprasad and Prakash, 2003). If these project-based activities are managed effectively, the CL can be used to reduce project time, improve quality and client satisfaction (Love *et al.*, 2003).

However to overcome the limitations in current industry practice, it is necessary that learning from a project is captured *while it is being executed*, and presented in a format that will facilitate its reuse during and after the project. The CL approach will:

- Provide mechanism that can be utilised at the project and post-project stages to identify learning;
- Facilitate the capture and reuse of the collective learning on a project by individual firms and teams involved in its delivery;
- Provide learning that can be utilised at the operational and maintenance stages of the asset's life cycle;
- Involve members of the supply chain in a collaborative effort to capture learning in tandem with project implementation, irrespective of the contract type used to procure the project from the basis for both ongoing and post-project evaluation.

4.6 Summary/conclusions

The work presented in this chapter was aimed at examining the extent to which CL can be used in the search for solutions in improving project delivery and achieving greater learning and integration. Collaboration through learning will provide the competitive edge that enables all the participants in a project development to prevail and grow. CL requires individual participants to adopt simplified, standardised solutions based on common framework and architecture. It is evident that the industry stands to reap many benefits collaborating in this way. The benefits offered by this approach, has the potential to significantly improve the quality and project performance. Some of these benefits have been discussed in detail in (Udeaja *et al.*, 2008) and are summarised here. They are as follows:

- Construction project team will benefit through the shared experiences that are captured as part of the learning on key events, which can have both short- and long-term values.
- Other project teams in an organisation can use the learning captured from previous/similar projects to deal with problems they encounter in another project.
- In the longer term, client will benefit from the increased certainty with which construction organisations can predict project outcomes.
- Improved project management, as supply chain members would work more collaboratively and share lessons learnt on construction projects.
- Project teams will benefit from an enhanced knowledge base as much learning that is presently not documented can be captured and reused.
- Provide knowledge that can be utilised at the operational and maintenance stages of the asset's life cycle.
- Involve members of the supply chain in a collaborative effort to capture learning in tandem with project implementation, irrespective of the contract type used to procure the project from the basis for both ongoing and post-project evaluation.

This chapter has proposed ways in which CL can facilitate, promote, enhance and support learning from a project to be used in the same project or future projects. In the current dynamic environments, the potential of collaborative team to enhancing learning can be even more important. For example, when project team member are faced with making quick decisions, CL approach can provide efficient and effective capabilities for capturing learning. Using this captured learning, team members can easily rectify the mistakes they made in a previous task with project development.

This chapter has introduced CL from a general perspective, which examined how the approach has been conceptualised in all industries and, in particular, the construction industry. The work identified that the approach of capturing learning in a project environment is quite novel and a fundamental departure from the current classical approach of capturing learning within an organisational setting. Furthermore, a review of contemporary related works were undertaken to identify how researchers have conceptualised this approach. The review unearthed interesting concepts, but it showed that not much work has been undertaken in the area of construction project procurement. This chartered a course for how CL can be implemented in construction.

To conclude, this chapter has described ways in which innovative project procurement methods can enhance the project team's activities by being better able to leverage learning from projects. Ultimately, improvements in the project procurement as a result of the CL approach can reduce the construction period and reduce the cost of projects.

5 Methodology for Live Knowledge Capture and Reuse of Project Knowledge

As mentioned in Chapter 3, case studies were conducted to explore and obtain deeper insights into the current approaches in the construction industry for the capture of reusable project knowledge, end-users' requirements for knowledge capture and reuse, the various types of reusable project knowledge and learning situations. This chapter presents the findings and analysis of the findings of the case studies, as well as the development of the methodology for the 'live' capture and reuse of project knowledge.

5.1 Background of case study companies

Six case studies were undertaken, involving semi-structured interviews with eighteen representatives of the six companies whose positions ranged from Group Knowledge Manager to Company Partner. The six case study companies were partners of the CAPRIKON (Capture and Reuse of Project Knowledge in Construction) research project, which forms the basis for this book. Background information on the companies is presented in Table 5.1.

These companies are different in terms of their business nature and size, and play different roles in a construction project. This helped to prevent bias and ensure that a variety of perspectives were obtained from the case studies.

5.2 Findings from the case studies

The findings from the case studies represent the collective views of the companies involved in the areas investigated, where significant overlaps of information were observed. These are combined and structured into the following subheadings:

- Types of reusable project knowledge;
- Learning situations;
- End-users' requirements for knowledge capture and reuse;
- Current approach for knowledge capture.

Table 5.1   Background of case study companies

Company name	Positions of interviewees	Company background	Number of employees	Annual revenue (£)
A	Partner, Associates and IT Manager	Design Consultant	80	£4.3M
B	Managing Director (Design), IT Manager, Systems Manager and Procurement Manager	Design Consultant, Developer and Contractor	850	£250M
C	Group Knowledge Manager and Knowledge Researcher	Engineering Consultant	7000	£403M
D	Group Knowledge Manager, Associate Director and Head of R&D	Management Consultant	1200	£61M
E	Director of Business Development, Senior Account Manager and Customer Support Staff	Project Extranet Service Provider	31	£2M
F	Knowledge Manager	Water Company	18000	£1860M

The types of reusable project knowledge and learning situations have been presented in Chapter 3, so only the latter two items are discussed in this chapter.

5.2.1   End-users' requirements for knowledge capture and reuse

The main requirement for the development of the methodology is to facilitate the capture and access of project knowledge at any time (i.e. 'live') and at any place. The design of the methodology must reflect the fact that project team members are often pressed for time and not always collocated. The methodology must therefore allow individual project team member to share and access knowledge at any time without making it compulsory for all project team members to meet face-to-face for the purpose. The case study companies acknowledged that the aforementioned requirements are critical, and identified the following requirements:

(a) *Cost*: The general consensus among the case study companies was that the methodology used for the capture and reuse of the reusable project knowledge should not incur significant additional cost to the companies. Furthermore, Company A noted that the cost incurred should also be justifiable by the benefits brought about through the reuse of the knowledge captured.

(b) *Workload*: The companies emphasised that any methodology developed should not create significant additional workload to members of staff in view of their existing heavy workload. They added that the additional workload created should be integrated into existing job functions and be carried out within normal working hours. They contended that this is the key to minimising rejection and securing acceptance from the people involved for the successful introduction of any new practice into an organisation. They also pointed out that the additional workload might not be covered by the worker's current job description or employment contract.

(c) *Legal issues*: Some companies prohibit their staff and collaborating companies from disclosing the information and knowledge gained to other organisations that are not involved in the project. A solution is required to ensure that the sharing, capture and reuse of knowledge from a project is not in breach of copyright and the conditions of contract.

(d) *Accuracy*: Any methodology developed must be capable of capturing and representing the knowledge accurately.

(e) *Representation of knowledge*: The main requirements for knowledge representation are summarised as follows:

   o A standardised approach is required. The knowledge captured must be organised and represented in a logical and simple to understand way, and be readily accessible to others within the organisation.

   o Case studies or detailed explanation of the knowledge are to be provided and shared in a Web environment to help others to understand and hence reuse the knowledge. They suggested that this can be supplemented by video clips to capture the detailed explanation from the originator of the learning.

   o A short description should be prepared to give the reader basic background information about a knowledge item, and the characteristics of the project that are related to the context for the reuse of the knowledge.

   o The conditions for reusing the knowledge must be made clear to the users.

   o There is a need to establish a convenient means, such as people's personal profile and knowledge network aided by custom-designed IT-systems, for people to communicate with each other and share their knowledge.

Appendix D summarises the individual company's requirements on knowledge representation.

5.2.2 Analysis of the end-users' requirements for knowledge capture and reuse

A methodology for the 'live' capture and reuse of project knowledge can be developed based on the various requirements identified from the case studies. The main requirements identified cover: (1) cost and workload, (2) legal issues, (3) accuracy of knowledge captured, (4) representation of knowledge and (5) facilitating the capture and reuse of project knowledge as soon as possible once it is created or identified. A workshop consisting of the representatives of the case study companies and academics was conducted to assess the strategy to be adopted to address these requirements, the categorisation of reusable project knowledge, the format used for capturing and representing knowledge of a proposed IT tool and the viability of the proposed methodology prior to the development of an associated IT tool. The workshop revealed that the methodology is viable, and the feedback and the suggestions obtained were subsequently incorporated into the design of the format used for representing knowledge and the methodology. These details are explained in the appropriate sections of this chapter. The measures taken to address the requirements are as follows:

Cost and workload

There are three cost components of a knowledge management (KM) system that have to be managed and taken into consideration in the development of a KM system/methodology (Robinson *et al.*, 2004):

- The staff costs (KM team component) associated with the roles and skills required for knowledge transformation.
- The organisational or (re)organisational costs (KM process component) associated with core and supporting business processes enabled, affected or re-engineered.
- The KM infrastructure component costs associated with information and communications technologies (ICTs) (hardware and software), and the setting up or maintenance of people sharing networks, systems or techniques.

The following recommendations can help to reduce and prevent additional cost in the aforementioned cost components:

- To keep the staff cost low, the 'live' capture and reuse of reusable project knowledge methodology should avoid the need for additional staff and the creation of significant additional workload for existing staff. Cost and workload are in fact interwoven as Robinson *et al.* (2004) have shown that staff cost is associated with the role or workload for knowledge transformation. Apart from contradicting with the end-users' requirement that significant additional workload is not desired, it may also reveal contractual issues as the additional workload created may not be covered by the current job description of the members of staff. Therefore, to resolve this matter it is suggested that most, if not all, of the relevant tasks and additional workloads created are handled by ICT (i.e. through an application software).

- To reduce the organisational or (re)organisational costs, the methodology developed should be built on existing practice if possible (i.e. integrated into something that people already do, such as meetings and reviews) for the capture of knowledge. This can help to prevent significant additional costs due to the need to re-engineer the current processes, and the creation of additional workload.
- To reduce the KM infrastructure component costs, the application software developed as part of the 'live' knowledge capture and reuse methodology should be capable of running on the existing ICT systems and platforms which are commonly used by the construction organisations or are readily available in the market. Otherwise, it could lead to significant cost increase and render the plan to implement the system commercially unfeasible.

This requirement is further discussed in the section on 'Enabling Technologies and Techniques'.

Legal issues

To overcome the client's potential restriction on sharing information and knowledge with parties not involved in the project, the knowledge to be shared can be limited to those captured from the current project. The sharing of knowledge captured from other projects should be voluntary. An appropriate legal framework for 'live' knowledge capture and reuse needs to be developed and agreed between the project team members.

Accuracy

A validation mechanism is required to ensure that the knowledge entered is accurate, complete with all the details required in the specified format, important and reusable as a means to prevent knowledge overload. In Company F, in which the new knowledge captured has to be validated by a panel of experts before it is published on the company's intranet for reuse, it can be used as a reference.

Representation of knowledge

The case studies revealed that reusable project knowledge often exists as a mix of tacit and explicit knowledge. Therefore, concentrating on either capturing explicit knowledge through codification of the knowledge or building a network of people for sharing tacit knowledge will fall short for managing reusable project knowledge effectively. To address this problem, the methodology was designed to explicate project knowledge into explicit form as far as possible since it is easier to be shared and transferred for reuse. For the remaining tacit knowledge which is really difficult to be explicated, links (e.g. contact details) are provided to connect the author of a knowledge entry with those who need it for the sharing of that knowledge. The methodology may seem to incline towards a codification strategy, but it also caters for tacit–tacit exchanges. According to Hansen *et al.* (1999), this is the approach adopted by firms who have excelled in managing their knowledge. A standard format for representing the reusable project knowledge captured (i.e. one of the requirements identified) was proposed and subsequently validated in the workshop conducted.

The concept of a Project Knowledge File (PKF) is introduced which contains relevant project information and project knowledge that can be reused both during the execution (e.g. in subsequent phases) and after the completion of the project. The PKF covers:

(a) *Background information on the project*: These include project title, project location, project sector, type of project, start and completion dates, duration, companies involved and date on which the knowledge is captured (which is included as an attempt to address the knowledge obsolescence issue).
(b) *Abstract*: This is a short description of the knowledge captured.
(c) *Details*: This is the detailed explanation of the knowledge so as to help others to understand and hence reuse the knowledge. Video clips, diagrams and photographs can also be used to help explain the details about the knowledge, or to capture the tacit knowledge.
(d) *Conditions for reuse*: This spells out the condition(s) for reusing a particular knowledge entry.
(e) *Reference*: This contains the reference to other relevant knowledge captured in the system, project documents, publications (e.g. books and reports), websites, where further details may be obtained. A hyperlink to Web pages showing the contact details of the author (e.g. phone number, email and photo) to aid the transfer of tacit knowledge is also provided here.

There is a consensus between the findings from the literature review (Maier, 2002; Rollett, 2003) and case studies (i.e. Companies A and D) that knowledge has to be put into theme-specific categories to ease understanding and retrieval. The reusable project knowledge identified can be organised in the hierarchy as depicted in Figure 5.1.

In addition to the organisation of knowledge, a knowledge map and an index can be provided to give users an overview of the knowledge available as suggested by Maier (2002). This can be met through the creation of an index table as depicted in Table 5.2.

Further to the strategy that the methodology should be built on existing practice in the construction industry for the capture of reusable project knowledge, the next section focuses on the details and the capability of current practice in meeting the design requirements for the methodology identified.

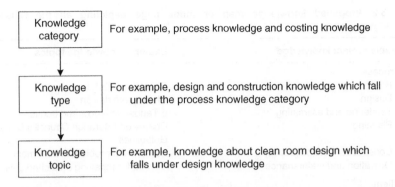

Figure 5.1   **The hierarchy for the organisation of reusable project knowledge captured**

5.2.3   *Current practice for the capture of reusable project knowledge*

The development of a methodology for 'live' capture of reusable project knowledge in construction requires an understanding of the current practice for capturing project knowledge and the capability of current practice to help facilitate it.

The case studies revealed that post project reviews (PPRs) are still the single most important KM tool to capture project knowledge. It also revealed that there was no single KM technique or technology that could meet all the requirements for knowledge capture in construction organisations. A very pragmatic approach was therefore adopted by the construction organisations to capture and reuse project knowledge with a combination of KM techniques and technologies used. The details of the findings on each of the current methods for capturing reusable project knowledge are presented below:

Post Project Review

The scope of knowledge to be captured in PPR is very wide as it covers almost all types of project knowledge. Company D conducts PPR within 6 months of the completion of its projects. PPR was chaired by project managers, and a report recording what went well and lessons learned was then prepared. The findings would be assessed to identify the learning which was reusable in other projects. PPR was made mandatory through the ISO 9000 system implemented by Company C. The PPR report was shared directly through its intranet and indirectly through its communities of practice (COPs). Company B's PPRs were conducted within 1–3 months of the end of a project, and took half or one full day depending on the complexity of the project and issues that emerged. The key learning was captured in point forms (i.e. do's and don'ts) which the company acknowledged as lacking in detail.

Table 5.2 Proposed knowledge map or index page depicting all the reusable project knowledge captured

Reusable project knowledge	Example knowledge topics
1. Process	
• Briefing	Checklist for briefing
• Design	Clean room design
• Tendering and estimating	e-Tendering – do's and don'ts
• Planning	Charnwood Borough Council's Development Guidelines
• Construction and buildability	Damp-proof problem in basement
• Operation and maintenance	Service for repairing scratched glass
2. Client	
• Clients' requirements	Company M's special requirements
• Client organisations' internal procedures	BAA's special internal procedures
• Background knowledge about client's business	Company K's business plan for 2006/2007
3. Cost	
• Cost of alternative forms of construction	Reduce waste with 'Dry Silo Mortar'
• WLC	How to reduce cost for lighting
4. Legal/statutory	
• Health and safety	Some notes about Part-L Regulations
• Changes in regulatory requirements	Introduction of some new EU standards into the UK
• Contract	PFI contract – important points
5. Technical details	
• Standard design details	Hospital R's standard design details
• Specifications	Specifications for the design of air-conditioning system for an operation room
• Method statements	Method statements for clean room construction
6. Performance of suppliers and KPIs	List of approved suppliers and their performance
7. Who knows what	Expert of siphon drainage
8. Others	
• Risk management	Importance of careful selection of subcontractors
• Team working	How we did it in 'City Walk' project – our experience
• Project management	Some do's and don'ts

Adaptations to PPRs were made by Companies A and F to overcome the shortcomings related to time constraint and the loss of knowledge due to lapse of time in capturing the knowledge. The companies conducted technical reviews at key stages during the course of the project to capture the learning in addition to PPR. In Company A, the learning was first documented by the project leader, and then reviewed and disseminated by the quality manager. In Company F, there were three phases of project review. For pre-project reviews, the project team members were required to gather information from other members of staff on the potential problems, issues, etc. that were likely to be encountered before the start of work on the project. Reviews were also conducted at predetermined milestones, such as between the design and construction stages, to capture the learning. This helped to reduce the time pressure on capturing the learning after the completion of a project and before people lose the drive as 'learning may no longer be seen as important after the event is over'.

Custom-designed software

Three case study companies used custom-designed software for the capture and reuse of explicit project knowledge. In Company D, custom-designed software was used to capture and analyse the cost information in order to identify the major elements that drive the cost of a particular type of building. Detailed and accurate estimates for a project could be provided by the software application based on basic information such as type of building, gross floor area and location of project. Custom-designed software was also used in Company B for the capture of the knowledge on performance of suppliers. Users can search for suitable suppliers based on the type of work (e.g. subcontractor for piping work) and geographical area with a view of their past performance. This facilitates better selection of suppliers for future projects. Company F created a knowledge base and provided each of its members of staff and suppliers with a login name and password to access and contribute their suggestions on how to improve current working procedure, and the resultant time and cost savings into the knowledge base. The suggestions were reviewed by the company's panel of experts before it was shared in the knowledge base, and the members of staff and suppliers were rewarded based on the financial impact of the improvement suggested.

Groupware

Groupware referred to were Lotus Notes™ used by Companies B, D and F, and Company C's custom-designed tool to support its CoPs. Details of how groupware can facilitate the capture and reuse of reusable project knowledge were discussed in Chapter 3.

Project extranets

The 'Workflow and Approval Process' module of Company E's extranet allows for the approval routes for all the items on the extranet to be specified. The documents can either be Approved, Un-approved, Rejected or Under-review. Although not purposely designed for KM, this feature can facilitate the peer-review process (i.e. validation) for the documented learning in order to seek suggestions for improvement before it is formally tagged as 'best practices' or 'lessons learned'. In Chapter 3, details of how project extranets can facilitate the capture and reuse of reusable project knowledge were briefly discussed.

Communities of Practice

Two types of CoPs were identified from the case studies: the conventional CoPs without the aid of ICT and the CoPs aided by ICT such as intranet and groupware. Company B had nine CoPs for nine disciplines: estimating, design management, admin secretarial and so forth, which fell into the former category. Knowledge was shared through people-to-people interactions within and across Company B's CoPs.

In Company C, its CoPs were called skills networks and were aided by ICT. The company assisted the setting up of emergent CoPs by providing them with in-house developed groupware to support group interactions and communication. To advocate CoPs, registration for joining was not required and there was no restriction over the number of CoPs that one can participate. In many instances, people belong to more than one CoP. Details of how groupware can facilitate the capture and reuse of reusable project knowledge is presented in Chapter 3.

Recruitment

Two companies (i.e. Companies C and F) stated that recruitment was used to capture the knowledge which is not available within the company. To achieve this, a detailed selection procedure was established in Company F to ensure that the recruits have very good command of the required knowledge and skills. Furthermore, the candidates would also be assessed on their willingness to share knowledge.

Forums

There were two types of forums identified from the case studies: the conventional and IT-aided. In Company A, a conventional forum is conducted on face-to-face basis at monthly intervals for the senior partners to share their knowledge with the associates of the company. The company argued that this allows the tacit knowledge residing in the head of the partners to be shared with and captured by others in the company. Company C's online forum allows members of staff to post questions and request for assistance from colleagues with the knowledge across the company's intranet. The online forum is a very powerful tool in locating and sharing knowledge, particularly when there is no formal record of 'who knows what' in a company.

Documentation of knowledge

There were attempts in four out of the six case study companies to document their design knowledge and best practices. For the design knowledge, a handbook which spells out the standard procedures to be followed in design, key design issues need to be paid attention (e.g. health and safety), the forms to be filled and the reference or Web links to relevant information, was created and circulated within Company A. The handbook was accessible through the intranet. Company C created feedback notes to capture the industry's best practices and lessons learned. It is also aimed at investigating how emergent issues are influencing the company's current practice and to provide suggestions on how to deal with the issues. Feedback notes were written in a standard format and were subject to peer review for validation before they could be shared in the company's intranet. Company D prepared case studies for each of the projects to record the roles of the company in the project, relevant background knowledge, programme and uniqueness of the particular project. The case studies prepared were accessible through intranet. For Company F, a knowledge base was created for the capture of knowledge in the format specified by the custom-designed software.

Expert directory

Expert directory refers to the Personal Profile, Divisionary Directory and staff appraisal report of Companies C, F and B respectively for the capture of the knowledge on 'who knows what'. This knowledge covers details such as the skills, experience, expertise, contact details and job function of company staff. It is crucial in facilitating the connection of people with the right knowledge to the people who need the knowledge, particularly the more tacit knowledge which is notoriously difficult to codify. 'Personal Profile' was Company C's intranet-based staff profiling system for capturing this knowledge. A standard procedure was established to ensure that members of staff keep their personal profiles up to date. Company B conducted staff appraisals at fixed intervals (normally annually) to capture this knowledge. However, the knowledge was recorded in paper form and there was no established means for other staff to access and hence to benefit from reusing this knowledge. Company F's Web-based 'Divisionary Directory' was very similar to the Company C's 'Personal Profile'. The system facilitated the identification of the right people with the right knowledge by name and keywords (e.g. the type of expertise required) for knowledge sharing.

Research and development team

The Research and development (R&D) team within Company D was established to seek room for improvement and encourages members of staff to suggest new topics for research. The research carried out could be regarded as one of the practices to acquire new knowledge within the company. The team provided brochures on the outcome of its research, and made presentations to the various branches within the company to disseminate this knowledge.

Team meetings, road shows, presentations and workshops

In Company D, road shows, presentations and workshops were held periodically and project teams from different sectors had monthly meetings to share knowledge. Although some knowledge was codified and made available in the documents disseminated at the meetings, most of the knowledge shared was captured in the heads of the people who attended the meetings.

Training

Companies A, D and F identified training as a practice for the capture of project knowledge. Training was provided to members of staff at fixed intervals (e.g. every 3 weeks) covering a range of topics. In Company A, training sessions were integrated with lunchtime CPD (Continuous Professional Development) sessions. Documented knowledge was disseminated in Company D's training, and tacit knowledge could be shared through the interaction between the trainers and trainees. External knowledge might also be captured as external experts were invited by Company A to give presentations.

Knowledge teams

Company D set up knowledge teams, which were led by key persons or experts in the respective business areas, to identify and capture the knowledge imperative to their fields. The company argued that this would allow other teams to tap into the knowledge captured for reuse in their respective fields. In addition, the knowledge gained could also be reused for training purposes.

Collaboration with other companies

Company D collaborates with other companies including its competitors in research and construction projects for the sharing of knowledge and information on benchmarking and best practices. Some of the knowledge was captured through observation and attempts made to replicate or innovate based on others' practices. Company F encourages and rewards its suppliers for contributing useful knowledge into its knowledge base. In addition, Company B also acknowledged the imperative of capturing knowledge through collaboration with other companies.

Preparation of the standard reusable details

Company B's project teams conducted special sessions to identify areas where standard details on design and specifications can be created, and to identify existing standard details for reuse. The standard details were created in electronic form and were made available to the team for reuse in other similar projects with the same client, or even for projects with other clients. This helped avoid the reinvention of the wheel and the need to start from scratch for each of the new projects. It had indirectly led to the savings both in terms of time and costs.

Reassignment of people

The study revealed that reassignment of people was the most direct method used by the case study companies to reuse the knowledge captured from one project in another project. This is particularly for the tacit knowledge which is notoriously difficult to capture. Besides reassigning people to other projects, Company C allocated members of staff who were experienced in similar type of project to provide assistance to other project teams. Company B also moved its members of staff from one discipline to work for a short period of time in another discipline. The company contended that bringing people from different disciplines together could help people to understand where the bits and pieces of knowledge were being stored within the company. In addition, the company also felt that more ideas could be generated through the interactions of people from different disciplines.

External sources of knowledge

Companies A and B identified tapping knowledge from external sources as one of the practices for the capture of reusable project knowledge. Company B noted that some of the project knowledge could be obtained by subscribing to the relevant service providers, such as Whole Life Cost Forum (www.wlcf.org.uk) for the knowledge on whole life costing and Building Cost Information Service (http://www.bcis.co.uk) for knowledge on building costs. According to Company A, the publications of government departments and other professional organisations such as the GLC (Greater London Council) Detailing for Building Construction and Architect Metric Handbook (a design guide) and product presentations by manufacturers or suppliers, which cover knowledge on a variety of problems, issues and their recommended solutions are amongst the other external sources of knowledge. The external sources of knowledge may lead to time and cost savings for the capture of knowledge, particularly those require a relatively long time for its capture such as knowledge on whole life costs (WLCs).

Succession management and mentoring

Only Companies A and C identified succession management and mentoring as the norm for the capture of reusable project knowledge respectively. Company A's succession management covers the identification of young architects to learn (or capture) the specialist design knowledge for pharmaceutical facilities from the experts. Company D's mentoring closely resembled the practice outlined in the existing literature, where junior staff were assisted in their work by attaching them to a mentor.

5.2.4   Analysis of current practice for the capture of reusable project knowledge

Various KM techniques and technologies were being used by the case study companies for the capture of reusable project knowledge (see Table 5.3). ICT was found to play a significant role in facilitating the 'live' capture and reuse of project knowledge. All of the KM techniques (e.g. CoPs and forum) and technologies that can partially satisfy the requirements identified for facilitating 'live' capture and reuse of project knowledge were either aided by ICT or are ICT tools themselves. The companies' current approaches to capturing reusable project knowledge are summarised in Table 5.3.

The shortcomings of current approaches in terms of the capability to facilitate the 'live' capture and reuse of project knowledge are discussed below:

- PPRs
  PPRs are normally time consuming and slow. The time lapse between the discovery and creation, and the capture and sharing of knowledge leads to the loss of important insights (Kamara *et al.*, 2003) and hence fails to facilitate the 'live' capture of project knowledge. Two other major shortcomings of current PPR practice were identified in the case studies: first, in three out of the five cases, the learning captured was not being shared effectively and there was no established way to locate the knowledge embedded in reports for reuse. Secondly, the current practice of distilling the key learning captured in PPR into point form is too brief for understanding the context and for efficient sharing of the knowledge captured.

  However, despite the inability to facilitate the 'live' capture of project knowledge, PPR is important for capturing the collective learning of the different parties involved in a project. Project reviews can be made more useful by shortening the interval between reviews (i.e. increasing the frequency of such project reviews). This can help to reduce the knowledge loss problem. In fact, the practice of Companies A and F to conduct project reviews at each of the key project stages for capturing project knowledge is a better alternative than PPR. This can even be further extended to capturing project knowledge at the routine weekly or bi-weekly project meetings, which probably represents the shortest interval possible for the capture of collective learning from a project team. Therefore, it is recommended that project reviews should be made part of the methodology for 'live' capture and reuse of project knowledge.

Table 5.3   KM techniques and technologies adopted by the case study companies

KM technique	Facilitating live capture and reuse of knowledge?	Case study companies using this practice for knowledge capture					
		A	B	C	D	E	F
PPRs	No	√	√	√	√	√	√
CoPs	Partially		√	√			√
Documentation of knowledge	Partially	√		√	√		
Training	No	√			√		√
Forum	Partially	√		√			
Recruitment	No			√			√
External source of knowledge	No	√	√				
Reassignment of people	No		√	√			
Research collaboration	No				√		√
Partnership-like arrangements	No					√	
Preparation of standard reusable details	No		√				
Research and development	No				√		
Team meetings, road shows, presentations and workshops	No				√		
Knowledge team	No				√		
Succession management and mentoring	No	√					
KM technology		**A**	**B**	**C**	**D**	**E**	**F**
Groupware	Partially		√	√	√		√
Custom-designed software	Partially		√		√		√
Expert directory	Partially		√	√			√
Project extranet	Partially			√			

- CoPs, groupware and forums

Without the aid of ICT, it was found that the conventional CoPs and forums fail to facilitate the 'live' sharing of project knowledge across geographical dispersed offices. Company A's practice to restrict the participation in its forum to senior staff made it impossible for others to directly benefit from the practice.

The online CoPs and forums have overcome the geographical constraints for sharing knowledge through the use of groupware and other custom-designed software. In addition, the knowledge shared and the threads of correspondences can be archived or saved by the ICT applications. This allows knowledge to be retrieved and reused in the future.

However, online CoPs and forums still fall short in meeting all the end-user requirements due to their passive nature. This is because if a question is not asked in the online CoPs or forum, the knowledge pertaining to the question is less likely to be shared. A more proactive approach is required.

In addition, there was no standard format created to represent the knowledge shared in the groupware used by the case study companies, which is one of the end-users' requirements identified earlier in this book. Furthermore, developing the methodology for 'live' capture and reuse of project knowledge in a groupware can be a very expensive option due to the licensing fees, etc. required. Therefore, groupware (as currently used in industry) is not considered as a suitable option for this purpose.

- Recruitment

It was observed that recruitment was used primarily for filling gaps in the case study companies' existing knowledge base rather than as a practice for the capture of knowledge from its ongoing projects. Other than this, it is a lengthy process undermined by the difficulties in finding and assessing experts with the required knowledge (Harman and Brelade, 2000) and the scarcity of experts (Maier, 2002).

- Training, team meetings, road shows, presentations and workshops

The time lapse between the capture of knowledge from a project to the sharing of the knowledge through these knowledge sharing mechanisms also suggests that they do not adequately facilitate the 'live' capture and reuse of project knowledge. Furthermore, the scope of knowledge available for sharing through the aforementioned practices are also constrained to those captured by the trainers and participants, and are normally topic-specific. Other than this, there was no established means observed from the case studies for the sharing of the knowledge captured with those who are not involved in the trainings, etc.

- Succession management and mentoring

Succession management and mentoring are time-consuming processes and hence cannot facilitate 'live' capture and reuse of project knowledge. Furthermore, succession management was only used to transfer a specific type of project knowledge in the case studies. According to Company A, its succession management was not very successful due to the reluctance of people to confine their learning to a specific area. Young architects prefer to be involved in different types of project instead of being restricted to one specific sector.

For mentoring, it is very efficient in the transfer of knowledge (particularly tacit knowledge). However, its efficiency is constrained by the number of protégés that the mentor can handle at any point in time, the distance between the mentor and potential protégé, issues related to cross-gender mentoring (Clawson and Kram, 1984), time constraints (Tabbron et al., 1997) and the ability of the mentor to transfer his/her knowledge to the protégé (Megginson, 2000).

- Documentation of knowledge

Companies A and D's checklist-based design handbook and case studies of project undertaken were criticised by their employees for lack of detail and reuse value. Companies C and F's practices (i.e. the creation of feedback notes which were accessible online and the maintenance of a knowledge base) were very mature and tested tools of documenting knowledge. However, there is no mechanism to ensure that the knowledge is captured 'live' or within a short time frame after its creation or generation. Such mechanisms, if created, may make a Web-based knowledge base the closest practice to meet the requirements for 'live' capture and reuse of project knowledge in construction. Furthermore, the knowledge captured by Company C's feedback notes is limited to that created or identified by the company while the views of other project team members are not captured.

- Partnership arrangement and research collaboration

Partnership arrangements and research collaboration are more a strategic arrangement for knowledge sharing rather than a practice for knowledge sharing by itself. Furthermore, these methods cannot guarantee that critical or key knowledge will be shared. This is because:

(a) The construction organisations collaborating in one project may actually be competing in another project (Kamara *et al.*, 2003);

(b) Corporate security restrictions imposed on posting of information/knowledge have further added to the problem (Ardichvili *et al.*, 2003). People have been indirectly discouraged from sharing their knowledge especially where the boundary of such restrictions is not made clear.

- Knowledge and R&D teams

The nature of work done by the knowledge and R&D teams seemed to be more relevant to knowledge creation and innovation than the capture of reusable project knowledge. Furthermore, the establishment of the knowledge and R&D teams entails additional resources which do not meet the requirements that significant additional workload or cost is undesirable.

- Preparation of the standard reusable details

It must be noted that this practice is probably only economically viable for companies with a high proportion of similar projects. In the case of Company B, this was justified by the fact that 80% of its projects are from 30 key clients. Furthermore, for people other than the creator of the documents or drawings, the reuse may pose some problems as the rationale for the design and changes made might not always be clear to them.

- Reassignment of people

The success of reassignment of people for knowledge capture and reuse depends heavily on: (1) the staff turnover rate (Kamara *et al.*, 2003), and (2) the individual's ability to capture the learning from his/her previous project and then reuse the knowledge in another project or share the knowledge with others. The fact that people are only reassigned to another project after the completion of existing project also suggests that this practice does not facilitate 'live' capture and reuse of project knowledge.

● External sources of knowledge

The external sources of knowledge may lead to time and cost savings for the capture of knowledge, particularly those that require a relatively long time to capture (such as knowledge on WLCs). However, what the companies obtained from the external sources was general project knowledge, rather than detailed reusable project knowledge. This is also not a practice for 'live' capture of project knowledge.

● Project extranets

Currently, the role of project extranets is more significant in the sharing of documented or explicit knowledge (such as the reusable project documents) rather than tacit knowledge. In addition, there is no specific template or mechanism specifically designed for the capture of project knowledge. However, project extranet can be a suitable medium to facilitate 'live' sharing of information and knowledge than intranet. This is because it can provide access to different organisations involved in a project for the purpose of capturing and accessing reusable project knowledge. By comparison, an intranet's access is restricted to a single organisation only.

● Expert directory

Web-based expert directory helps to facilitate the 'live' identification of the right people with the right knowledge, which in turn assists in the 'live' sharing and reuse of project knowledge. However, expert directory captures only the knowledge on 'who knows what', and is not appropriate for the capture and creation of other types of knowledge.

● Custom-designed software

Tan *et al.* (2004) have identified various types of reusable project knowledge in construction, which need to be managed. Custom-designed software systems used for the capture of project knowledge were, however, narrow in scope and focused on specific types of project knowledge only. For instance, Companies B and D's custom-designed software targeted only costing knowledge and knowledge about the performance of suppliers. It was noticed that the Web-based nature of the custom-designed software could greatly enhance the sharing and reuse of knowledge. This emphasises the importance of Web-based technology in facilitating the 'live' capture and reuse of project knowledge.

Overall, the findings from the case studies revealed that although there are various KM techniques and technologies available, none of these represents a complete solution. The findings further revealed that both KM techniques and technologies have their strengths and shortcomings, and may in fact complement each other. Therefore, a combination of KM techniques and technologies is more likely to meet the various end-user requirements for the development of a methodology for 'live' capture and reuse of project knowledge as outlined earlier in the book. This is further explored in the next section.

Enabling technologies and techniques

The essence of the 'live' capture and reuse of project knowledge methodology lies in allowing users at different locations to enter and access the knowledge captured in real-time. The strength of Web-based KM technologies (such as groupware, expert directories and knowledge bases) is an integral element of the methodology for the 'live' capture and reuse of project knowledge. This is due to their capability to connect distant offices together, provide fast access to and location of knowledge captured, facilitate sharing of knowledge and provide huge knowledge storage space. Among the KM technologies available, a Web-based knowledge base seems to be the current practice closest to meeting the requirements identified. The reasons are as follows:

- *No significant additional cost*: The pervasive use of intranets by companies to connect their offices together has laid the necessary foundation for implementing a Web-based knowledge base. A Web-based knowledge base can run on the existing intranet/Internet systems and platforms commonly used by most of the construction organisations. This eliminates the chances of incurring significant additional cost for the implementation of the methodology.
- *No significant additional workload created*: The only requirement is the need to enter project knowledge into the knowledge base.
- *Accuracy of knowledge ensured*: A mechanism can be built into the knowledge base for monitoring the validation of knowledge submitted as a means of ensuring its accuracy.
- *Allowing a standard format for representing project knowledge to be specified*: Another built-in mechanism can be created to ensure that project knowledge is entered in accordance with the format developed.
- It can provide the necessary platform for access to and sharing of knowledge which is captured in the form of video clips and other formats of multimedia files.
- It may be used in conjunction with other Web-based applications (e.g. groupware and video conferencing tools) to enhance the sharing of knowledge, particularly the tacit knowledge.
- It can be integrated with 'skills yellow pages' which captures the knowledge about 'who knows what' within an organisation. This helps in the location of the author of the knowledge and the people with the right knowledge.

The case studies had recognised that the methodology must be designed to capture knowledge from both individuals and in a group setting so that useful knowledge generated in various learning situations will not be overlooked. Capturing knowledge in a group setting also helps to ensure a more holistic and more complete set of knowledge is captured than through individuals alone. Therefore, in addition to allowing individuals to submit knowledge into a knowledge base, the Web-based knowledge base will be supplemented by PPR and interim project meetings/reviews. Knowledge will be discussed and recorded, and subsequently entered into the Web-based knowledge base. PPR and project meetings/reviews were chosen for capturing knowledge in a group setting since this option creates less additional work. This is because conducting PPR has already become part of the quality system requirements of many construction companies, whereas conducting routine project meetings is a must-do task for all construction projects.

5.3   Structure of the 'live' capture and reuse of project knowledge methodology

A methodology for the 'live' capture and reuse of project knowledge has been developed based on the findings from the case studies and literature review. The methodology comprises:

- *A Web-based knowledge base*: This is where the Project Knowledge File (PKF) of a project is stored. A PKF contains relevant project information and project knowledge that can be reused both during the execution (e.g. in subsequent phases) and after the completion of the project. A PKF is similar to the Health and Safety File (HSF) under the Construction (Design and Management) (CDM) Regulations in the United Kingdom. The difference being that HSF is a project record which focuses on health and safety (HSE, 1997), whereas the PKF targets reusable project knowledge.
- *A Project Knowledge Manager (PKM)*: This is a role, normally charged to a project manager or other designated person, to manage the knowledge base (i.e. the development of a PKF for a project) and the Integrated Workflow System (IWS).
- *An IWS*: This delineates, executes and monitors the mechanism for the capture, validation and dissemination of the project knowledge captured. A PKM may configure the IWS to suit individual requirements of the project. See Section 5.3.1 for details.

The proposed methodology is designed to capture reusable project knowledge generated from the various learning situations once the knowledge is created or identified (i.e. 'live') through project reviews/meetings (i.e. group of people) and individuals (see Figure 5.2). Users have to enter reusable knowledge in accordance with the format specified. The knowledge captured from individuals needs to go through a

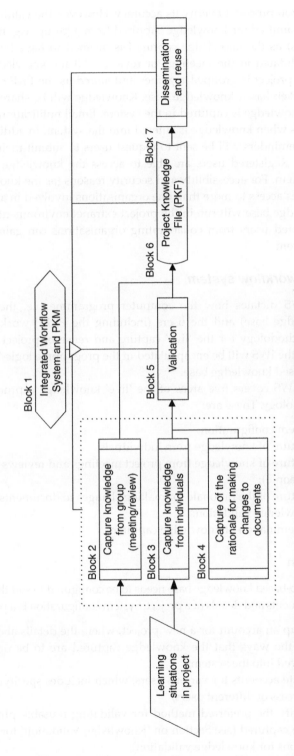

Figure 5.2 Methodology for the 'live' capture and reuse of project knowledge in construction

validation process to verify its accuracy. However, the validation process can be omitted for knowledge captured from a group (e.g. meetings and reviews) as the knowledge captured is deemed to have been reviewed and validated in the meetings or reviews. All the knowledge captured from a project is grouped together and stored as the PKF of the project in the Web-based knowledge base. Knowledge will be shared 'live' soon after knowledge is captured in the system. Email notification will be sent to users when knowledge is entered into the system. In addition, routine email reminders will be sent to request users to submit their knowledge entries. Registered users are able to access the knowledge captured in the system. For accessibility and security reasons (as the knowledge base provides access to more than one organisations involved in a project), the knowledge base will run in the project extranet environment where only designated users from collaborating organisations can gain access into the system.

5.3.1   Integrated workflow system

The IWS dictates how the computer programme (i.e. the Web-based knowledge base) and the users (including the PKM) work together in the methodology for the 'live' capture and reuse of project knowledge. Part of the IWS will be encapsulated in the programme logic/codes of the Web-based knowledge base.

The IWS covers five areas of the 'live' knowledge capture and reuse methodology. These are:

- System configuration;
- Capture of knowledge from individuals;
- Capture of knowledge from project meetings and reviews (i.e. a group of people);
- Capture of the rationale for making changes to documents;
- Knowledge validation;
- Dissemination of knowledge captured.

System configuration

The Web-based knowledge base needs to be configured to suit the individual requirements and details of a project. System configuration is a process to:

- Set up an account for a new project, where the details about a project and the ways that the knowledge captured are to be organised are entered into the system;
- Create accounts for various users, which includes specifying the level of access of different users;
- Specify the preferred method for validating reusable project knowledge captured (see Section on 'Knowledge Validation' for the various options for knowledge validation);

- Configure whether individual user would like to receive email notification when knowledge is entered into the system and when the status of the knowledge entered has been updated;
- Configure the system for sending out email reminders to the PKM to include knowledge capture in the agenda of coming project meeting or review;
- Configure the system for sending out routine email reminders to users to add new knowledge into the system.

This is a one-off process. It helps to avoid the need for re-entering similar information in the knowledge capture process. It is also needed in order for some features of the system to function, such as the automated email notification when a new knowledge item is entered into the system.

Knowledge capture

This process indicates how reusable project knowledge can be captured 'live' from ongoing construction projects. Three sources of reusable project knowledge were identified from the case studies, that is the individuals involved in a project, project meetings/reviews and the rationale for making changes to documents such as drawings.

- *Capture of knowledge from project meetings/reviews*
  The PKM will be responsible for including the capture of reusable project knowledge as an agenda item in the routine project meetings/reviews. The system will send the PKM an email reminder for this purpose. During the meetings/reviews, the learning captured since the previous meeting/review is discussed and the details agreed. If the system is accessible during the meeting/review, the designated person (who is normally the PKM) may enter the approved knowledge directly into the knowledge base in the specified format. Otherwise, the designated person may transfer the record into the system at a later time.
- *Capture of knowledge from individuals*
  All knowledge workers involved in the project will be assigned a login name and password to access the system. This allows them to enter their knowledge into the software tool once knowledge is created or identified (i.e. 'live'), or at anytime which is convenient to them. The system will send routine email reminders to the users who have opted to receive them.
- *Capture of the rationale for making changes to documents*
  The findings of the case studies revealed that the rationale for making changes to project documents (such as engineering drawings) is important reusable project knowledge. The system will provide a summary of the number of changes made to the project documents which the PKM can check at predetermined intervals. If there is a project document for which the number of changes made to it is well above average, a procedure for the capture of the rationale for making the changes

to the document can be invoked by the PKM. The author of the project document will be requested to provide the necessary details to the system. Similar to the knowledge submitted by individuals, the rationale for changes made to documents will be subject to validation before it can be disseminated.

Knowledge validation

It is important that the knowledge captured in the course of a project is as accurate and reliable as possible. This requires that it is validated. The knowledge captured from a group (in meetings and reviews) is deemed to have been validated whereas the knowledge submitted by individuals may need to be validated prior to reuse. However, at the organisation's discretion, the validation process may be omitted for the knowledge submitted by their experts and very experienced staff. The validation mechanism is triggered once new knowledge is entered into the system by individuals. There are two knowledge validation routes:

(a) To validate the knowledge submitted in the routine meetings or reviews. The knowledge submitted by individuals since the last meeting/review will be discussed at the current meeting/review. The PKM will be responsible for deleting or removing the knowledge from the system if the knowledge is rejected, or updating the status of the knowledge from 'draft' to 'validated' knowledge in the system if the knowledge is approved.

(b) Online validation. All the project participants or designated people/experts will be requested by the system to take part in the process within the predetermined deadline. The system will monitor progress and reminders will be sent if there is any delay in response on the part of the users. Four options for validating knowledge are provided. These include:
   (1) Semi-automated;
   (2) Rating-based;
   (3) Majority's opinion-based;
   (4) No validation required.

The key elements of the above options are provided below:

- *Online validation – comment-based*
   Users submit their comments on the draft knowledge and their opinions on whether the knowledge should be validated or not. After the predetermined deadline expires, the PKM reviews the comments and decides whether to validate the knowledge or not.
- *Online validation – rating-based*
   This is a variant of the comment-based option. In addition to allowing users to submit their comments, ratings (ranging from 1 to 5 stars) are

provided by the users for the draft knowledge. The PKM then decides whether or not to validate the knowledge based on the average rating and the comments made by other users for the draft knowledge. The PKM can also request that the content of a knowledge item to be revised. The comments given are accessible to others.

● *Online validation – majority's opinion-based*

Users select whether or not to validate the draft knowledge. The draft knowledge will be validated or removed from the PKF based on the majority's opinion after the predetermined period for review has expired.

● *Online validation – no validation required*

The system will bypass the validation mechanism. There is no distinction between draft knowledge and validated knowledge.

For the first three options, the author(s) of the knowledge will be informed about the status of the knowledge submitted (i.e. rejected or accepted) together with the rating and comments (if any) given by others.

Dissemination of knowledge captured

The 'live' sharing of project knowledge is achieved by proactively and automatically emailing users to notify them about the addition of a new knowledge item. The users should be notified of the changes made to the status of a knowledge item in the system. In addition, knowledge is also made available for access once it has been added into the system.

The methodology for the 'live' capture and reuse of project knowledge has been encapsulated into a software prototype called Capri.net. The next chapter presents the details of the automation of the software methodology.

6 The Capri.net System

This chapter presents the system architecture of the prototype application for the 'Live' capture and reuse of project knowledge methodology, and also covers the development, operation, testing and evaluation of the Capri.net prototype application. Two types of test – Acceptance Test and Entity-Life Histories (ELHs) Test – were first conducted on the prototype application. Evaluation of the prototype application was subsequently undertaken by a selection of industry practitioners who participated in the case studies described in Chapter 5. Based on the findings of the evaluation, the prototype application was further refined.

6.1 System architecture of prototype application

To automate the methodology for the 'Live' capture and reuse of project knowledge, a prototype application which consists of a Web-based knowledge base was developed. The system architecture of the prototype application is shown in Figure 6.1. The knowledge base will run in a project extranet environment which is only accessible to designated users from collaborating organisations. The database or the data layer is the core of the Web-based knowledge base where all the knowledge is stored. The application logic/code automates the Integrated Workflow System (IWS) and helps to reduce potential workload of the users in submitting and sharing reusable knowledge. A standard Web browser is used to interact with the knowledge base (i.e. for submitting or retrieving knowledge).

6.2 Development of the Web-based knowledge base

This section covers the selection of the most appropriate development environment and tool, and the development of the Web-based knowledge base using Web Information Systems Development Methodology (WISDM) proposed by Avison and Fitzgerald (2003). The development started with the design of the various user interfaces, followed by the creation of the databases and the writing of associated programme codes.

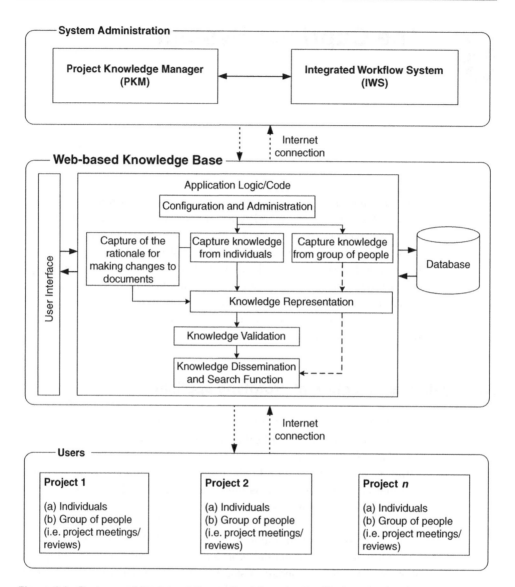

Figure 6.1   System architecture of the methodology for the 'live' capture and reuse of project knowledge in construction

However, there was also a degree of iteration in the development process. The details of the development process are described below.

6.2.1   Selection of development environment

The selection of a suitable development environment for the Web-based knowledge base impacts on the speed of development, and the cost of developing and running the end product. A number of options were available for the development of the Web-based knowledge base. However, the most appropriate ones were:

- The development of the prototype using Lotus QuickPlace (or similar application).
- The development of the prototype using a PHP and MySQL combination.
- The development of the prototype using a Microsoft™ ASP.NET 2.0 and Microsoft™ SQL Server Express 2005 combination.

The details of the suitability of these options are described in the following sections.:

The development of the prototype on the Lotus™ QuickPlace (or similar)

Lotus™ QuickPlace is a software application for running intranet/extranet services. Some companies use Lotus™ QuickPlace to create their Web-based database. However, this option was eliminated due to the high cost involved (e.g. the annual licensing fees for Lotus QuickPlace, which is over £10000 p.a.). In addition, this option was also discarded due to the need for a dedicated Web server to host or run the prototype application developed.

The PHP and MySQL combination

PHP is an open source scripting language used mainly for developing server side applications and websites, which include Web-based database (Wikipedia, 2006a). PHP is often combined with MySQL, which is a free SQL Database Management System, for the development of Web-based database (Wikipedia, 2006b).

The PHP engine and MySQL database server can be downloaded free-of-charge from the Internet. The PHP programme codes can be written using any word processor (e.g. Microsoft™ Windows's built-in Notepad). However, writing the PHP codes using a word processor is less efficient, slow and difficult to identify the errors in the codes. Using an integrated development environment (e.g. Zend Studio) for the development of PHP applications is advisable to address the aforementioned issues. However, this comes with a cost and offers less functions if compared to the Visual Web Developer (VWD) Express used for the development of ASP.NET 2.0 and Microsoft™ SQL Server Express 2005 applications. Furthermore, the PHP-MySQL option also has a longer learning curve than the ASP.NET 2.0 option.

ASP.NET 2.0 and Microsoft™ SQL Server Express 2005 combination

This combination is one of the latest Web-based database development technologies offered by Microsoft™. ASP.NET 2.0 is the equivalent of

PHP, but more powerful in terms of the range of functionalities offered. An integrated development environment (i.e. VWD Express for the development of ASP.NET 2.0 applications) is freely available from Microsoft™. In addition, VWD also comes with a free Microsoft™ SQL Server Express 2005 (i.e. the equivalent of MySQL). Compared to Zend Studio, used for developing PHP applications, VWD offers the following advantages:

(a) It offers a drag and drop feature for the creation of various controls on a Web Page, such as the user login, logout and forgotten password controls. Associated codes for the controls can be generated automatically by VWD. This helps to reduce the development time.
(b) It comes with a built-in security system. Different end-user roles with different access authentications can be easily created without the need for writing complicated programme codes. This again helps to accelerate the development of the Web-based knowledge base.
(c) It offers a fully integrated development environment. The management of the database, development of the programme codes and the debugging of the application can be done through VWD.

This option was chosen for the purpose of developing the Web-based knowledge base for the aforementioned reasons.

6.2.2   User interface and programme codes development

User interface (i.e. human–computer interface) design is critical in the development of the Web-based knowledge base. This is because it affects the user-friendliness of the system and also impinges on the design of the database structure. In the ASP.NET 2.0 environment, the development of user interfaces and programme codes are often carried out in parallel. This is because ASP.NET 2.0 uses a code-behind structure where the programme codes associated with the functions/features of a user interface are saved as part of the interface's source codes. This means that some of the features/functions in the user interface are not visible until the associated programme codes are written. This may slow down the progress of the prototype development as the user interface cannot be shown to the potential users for feedback before the associated programme codes are fully developed. Furthermore, at that stage, it would be too late to introduce any changes to the interface design due to the extensive rewriting of the programme codes or the redesign of the database structure required.

To address these issues, draft mock-up user interfaces were first designed using Microsoft™ Visio. Microsoft™ Visio allows mock-up user interfaces to be created quickly without needing the associated programme codes to be completed. The IWS and mock-up user interfaces were used to demonstrate graphically to the potential users in a mini-workshop (involving industry practitioners) about how the prototype application operates. This was done before the user interface and programme codes were developed (see Section 6.3 for details of the workshop and the feedback received). The user interfaces were then refined based on the feedback received. Subsequently, the working versions of the interface were developed using VWD. The user interfaces and associated programme codes were developed for the following tasks and associated challenges.

Capturing knowledge

Challenge: The design of the knowledge capture user interface and associated functions was geared towards minimising the need for re-entering duplicate information. This is critical in order to reduce the creation of additional workload to the users. A file upload function was also to be created for users to upload relevant documents and image files. Related to this, the programme codes written also must be able to prevent the accidental overwriting of existing file with a similar file name.

Solution: A number of dropdown menus were created on the user interface for capturing knowledge to avoid the need for re-entering information, such as project details, and different categories and types of knowledge (see Figure 6.2). These dropdown menus were linked to the respective tables in the database. They will be automatically updated with the changes made to the information in the database. Furthermore, the programme codes were written to automatically capture information such as the date of submitting

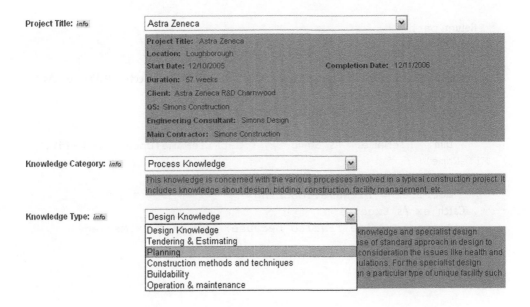

Figure 6.2   Dropdown menus for knowledge capture

a knowledge entry, the author's details, the calculation of duration of a project, etc. To prevent the accidental file overwriting problem, the programme codes were designed to examine whether there was an existing file with the similar file name in the system before a new file was uploaded (see Excerpt 6.1). If yes, the newly uploaded file will be saved with a different and unique file name.

Representing knowledge

Challenge: The details of a knowledge item are scattered across various tables in the database. The challenge is to retrieve the relevant information of a knowledge item such as the details of project and

```
Private Function GetFileName(ByVal filename As String) As String
  Dim i As Integer = 0
  Dim path As String = Server.MapPath("~/Documents/") & " ' " &
filename
  Dim fname As String = path.Substring(0, path.IndexOf("."))
  Dim ext As String = path.Substring(path.IndexOf("."))

  Do While File.Exists(path)
    i += 1
    path = fname & i.ToString() & ext
  Loop

  Return path
End Function


Protected Sub uploadButton_Click(ByVal sender As Object, ByVal e As
System.EventArgs) Handles uploadButton.Click
  If (DocFileUpload1.HasFile) Then
    Try
      Dim   fileName   As String   =   GetFileName(DocFileUpload1.
FileName)
      DocFileUpload1.SaveAs(fileName)
      docFile1 = fileName.ToString()
    Catch ex As Exception
      Response.Write("Failed because: <br/>" & ex.Message)
    End Try
  End If
End Sub
```

Excerpt 6.1   Codes for overcoming similar file name problem

author from the tables and display them on two user interfaces: the 'Summary' Page and 'Knowledge Details' Page. The programme codes must be written to pre-render the information into required formats prior to displaying them on the user interfaces.

For the 'Summary' Page, the main challenges include:

- Highlighting the latest knowledge entries in the system;
- The automatic creation of an abstract based on the details of a knowledge item. This is to eliminate the need for the author to prepare an abstract and to reduce the additional workload created;
- Highlighting the knowledge items which are tagged as 'Draft' or 'To Be Reviewed' for validation purpose.

For the 'Knowledge Details' Page, the main challenges are:

- The automatic rendering of URL in the knowledge details to a hyperlink;
- The automatic creation of download links to the uploaded files.

Solution: For the 'Summary' Page, the abstract of knowledge items were created by writing the programme code that extracts only part of the details of the latest knowledge items entered (see Excerpt 6.2). The programme code was also amended to retrieve only the list of knowledge items which were tagged as 'Draft' or 'To Be Reviewed' only onto the 'Summary' Page.

On the 'Knowledge Details' Page, all the contents sections with http:// will be rendered as a hyperlink. This was made possible by creating

```
CREATE PROCEDURE mysp_Get_LatestTop5

AS
   SELECT TOP (5) KnowledgeDetailsTable.KnowledgeID,
KnowledgeDetailsTable.KnowledgeTopic, ProjectDetailsTable.
ProjectTitle,
   SUBSTRING(KnowledgeDetailsTable.KnowledgeDetails, 1, 200) AS
Expr1, KnowledgeDetailsTable.DateEntered
   FROM KnowledgeDetailsTable INNER JOIN ProjectDetailsTable
ON KnowledgeDetailsTable.ProjectID = ProjectDetailsTable.
ProjectID
   ORDER BY KnowledgeDetailsTable.KnowledgeID DESC
   RETURN
```

Excerpt 6.2   Stored procedure for creating abstract of a knowledge item

programme codes that automatically identify these sections and then enclose them with the hyperlink's syntax before storing into the database (see Excerpt 6.3).

The download link to the uploaded file is essentially a hyperlink. It will navigate to the location of the uploaded file in the system when clicked. Codes are written to create the relevant download link automatically (see Excerpt 6.4).

Validating knowledge

Challenge: It must be noted that only the rating-based validation mechanism was built into the Web-based knowledge base as the proof of concept. First of all, there was a need for a mechanism to distinguish between the knowledge captured from individuals (which has to be validated) and the knowledge captured from meetings/reviews (which is deemed to have been validated during discussions at the meeting/review). A mechanism for managing the users' comments and ratings for a knowledge item was required. Furthermore, the Project Knowledge Manager (PKM) should be provided with an additional function for validating and deleting a rejected knowledge item from the system.

Solution: No specific page was created for the purpose of knowledge validation. The programme codes and features for knowledge validation function were built into and scattered in the user interfaces for capturing and representing knowledge (i.e. the 'Add Knowledge' and 'Knowledge Details' Page). On the 'Add Knowledge' Page, a

```
Dim knowledgeDetails As String = detailsTextBox.Text
Dim find As String = "(?<url>http://(?:[\w-]+\.)+[\w-]+(?:/[\w-
./?%&=]*)?)"
Dim Result As String = Regex.Replace(knowledgeDetails, find, "<a
                    href=""${url}"">${url}</a>")
e.Command.Parameters("@knowledgeDetails").Value = Result
```

Excerpt 6.3   Programme codes for hyperlink conversion

```
<asp:Label ID="Doc1Label" runat="server" Text='<%# Bind("Doc1") %>'></
asp:Label>

<asp:HyperLink ID="doc1DownloadLink" runat="server" NavigateUrl='<%#
Eval("Doc1", "~\Documents\{0}") %>' Text="Download"></asp:
HyperLink>
```

Excerpt 6.4   Programme codes for creating document download links automatically

dropdown list was provided for the users to specify the source of knowledge (i.e. individual or meeting/review (see Figure 6.3). If the source of the knowledge is 'individual', then the default status of that knowledge will be stored as 'Draft' in the database. If the source of knowledge is 'meetings/reviews', the default status of that knowledge will be stored as 'Validated'.

In the user interface for representing the details of a knowledge item (i.e. the 'Knowledge Details' Page), a text box and a dropdown list were provided for collecting users' comments and ratings for a 'draft' knowledge item respectively (see Figure 6.4). A database table was created for storing these information. Programme codes were then written to automatically retrieve the ratings given, calculate, update and display the average rating for the 'draft' knowledge item (see Excerpt 6.5).

A separate function, which is only visible and accessible by the PKM was also created at the bottom of the user interface for representing knowledge details (see Figure 6.5). The function allows the PKM to edit/change the status of a knowledge item in the main database from 'Draft' to 'Validated' based on the comments and average rating received. The changes made to the status of a knowledge item will then trigger the system to send out email notifications to the users.

Searching knowledge

Challenge: Two types of search function need to be created in the Web-based knowledge base: a simple Google™-like search function and

Captured From: info
Individual
Individual
Meeting/Review

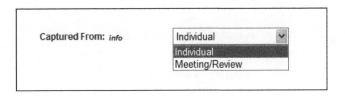

Figure 6.3 Dropdown menu for selecting the source of knowledge

Your Comment:
* Constructive comment is welcomed, whilst offensive comment is not allowed. Person who has given offensive comment may be barred from using this application.
Rating: 1 Star
Insert Cancel

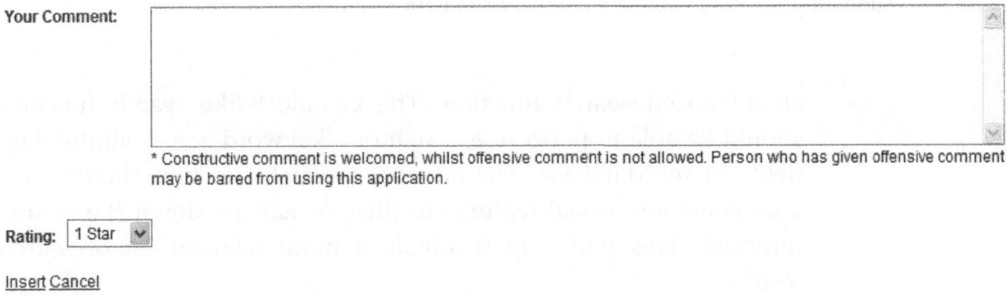

Figure 6.4 Function for entering comment and rating for a knowledge item

```
While rdr.Read()
  TotalRating += Convert.ToInt32(rdr("Rating"))
  If rdr("Rating") IsNot Nothing Then
    NumOfComment += 1
    Average = Convert.ToDouble(TotalRating / NumOfComment)
  End If
End While


If NumOfComment = 0 Then
  noCommentLabel.Text = "No comment received so far. Why
don't you be the 1st person to give your comment?"
  NumOfCommentsLabel.Text = "0"
Else
  NumOfCommentsLabel.Text = NumOfComment.ToString()
  noCommentLabel.Text = ""
End If


Dim AvgResult As String
Dim NFI As NumberFormatInfo
NFI = New CultureInfo("en-US", False).NumberFormat
AvgResult = Average.ToString("N", NFI)
AvgRatingLabel.Text = AvgResult.ToString()
```

Excerpt 6.5   Codes for calculating the average rating for a knowledge item

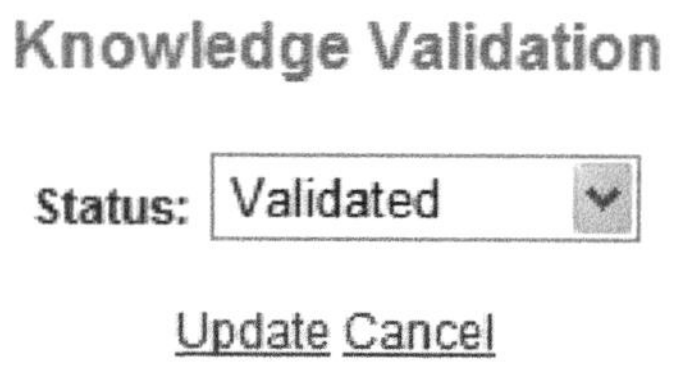

Knowledge Validation

Status: | Validated   ⌄ |

Update Cancel

Figure 6.5   Function for PKM to validate a knowledge item

an advanced search function. The Google™-like search function should be able to perform a search of a keyword across all the data fields in the database. The advanced search function should provide some additional features to filter or narrow down the results returned. This will help to obtain a more relevant set of search results.

Solution: Search textboxes were created for the Google™-like search function on the Index Page. A complex SQL query was written to compare

SelectCommand="SELECT KnowledgeDetailsTable.KnowledgeID,
KnowledgeDetailsTable.KnowledgeTopic, KnowledgeDetailsTable.userName,
KnowledgeDetailsTable.DateEntered, KnowledgeDetailsTable.CapturedFrom,
KnowledgeCategoryTable.KnowledgeCategory, KnowledgeTypeTable.KnowledgeType,
MemberInfo.email, ProjectDetailsTable.ProjectTitle, KnowledgeDetailsTable.status,
KnowledgeDetailsTable.KnowledgeDetails FROM KnowledgeDetailsTable INNER JOIN
KnowledgeCategoryTable ON KnowledgeDetailsTable.KnowledgeCategoryID =
KnowledgeCategoryTable.KnowledgeCategoryID INNER JOIN KnowledgeTypeTable ON
KnowledgeDetailsTable.KnowledgeTypeID = KnowledgeTypeTable.KnowledgeTypeID INNER
JOIN MemberInfo ON KnowledgeDetailsTable.userName = MemberInfo.userName INNER
JOIN ProjectDetailsTable ON KnowledgeDetailsTable.ProjectID =
ProjectDetailsTable.ProjectID WHERE (ProjectDetailsTable.ProjectTitle LIKE '%' +
@ProjectTitle + '%') AND (KnowledgeCategoryTable.KnowledgeCategory LIKE '%' +
@KnowledgeCategory + '%') AND (KnowledgeTypeTable.KnowledgeType LIKE '%' +
@KnowledgeType + '%') AND (KnowledgeDetailsTable.status LIKE '%' + @status + '%')
AND (KnowledgeDetailsTable.KnowledgeTopic LIKE '%' + @KnowledgeTopic + '%') AND
(KnowledgeDetailsTable.KnowledgeDetails LIKE '%' + @KnowledgeDetails + '%') ORDER
BY KnowledgeDetailsTable.DateEntered DESC"

Excerpt 6.6   SQL query for advanced search function

the text entered into the textbox with all the data in the main database. The results returned will include the knowledge items that contain the particular text, as well as the text which closely resemble the text searched. For example, if the word 'door' is searched, the results returned will include the knowledge items that contain the words 'indoor' and 'doors'.

An 'Advanced Search' Page was specifically created for the advanced search function. It can perform a search by combining a variety of terms (e.g. project title, knowledge category, knowledge type and keyword) to help construction a more detailed search. A number of new search criteria (e.g. project title and knowledge category) were created to allow the search results to be restricted to the knowledge entries, for instance, those that contain a particular keyword and fall under a particular knowledge category in a particular project (see Excerpt 6.6).

Configuring the system

Challenge: Certain information is required each time a knowledge item is entered into the system (e.g. project title, knowledge category and knowledge type). To avoid the need to re-enter similar information, a mechanism that allows such information to be entered once but retrievable over and over again was required. This feature was crucial to help reduce the additional workload created and to make the system as user-friendly as possible.

A mechanism was also required for:

- The PKM to configure when the system should send email reminders to him/her to include knowledge capture into the agenda of the coming project meeting/review.
- The PKM to configure the intervals at which the system should send email reminders requesting all users to add new knowledge into the system.

Solution: Before the system is ready for uploading reusable project knowledge, it needs to be configured. This includes adding information such as the project details, various categories and types of reusable project knowledge and user details. Various user interfaces were created for these information to be added into the database. These information were then linked to the various dropdown menus on the 'Add Knowledge' Page (see Section 6.2.2.1 and Figure 6.2 for details). This enables the users to click to select the required option from the dropdown menus without the need to re-enter the information. The various information items required and associated user interfaces created are depicted in Table 6.1.

The mechanism for the PKM to configure the intervals for sending email reminders to users was achieved through writing an email script for the purpose, and then use the Windows built-in 'Scheduled Tasks'

Table 6.1   User interfaces for capturing repetitive information

User interface	Information captured
'Add New Project' Page	Project details, i.e.: • Project title; • Project location; • Client; • Quantity surveyor; • Engineering consultant; • Architect; • Main contractor; • Start and completion dates; • Duration (automatically calculated and added by the system).
'Knowledge Category and Type' Page	• Various knowledge categories and associated definitions; • Various knowledge types and associated definitions.

function to execute the email script at the required intervals. The procedure is as follows:

(1)   Add a new scheduled task through the 'Scheduled Tasks' function.
(2)   Configure the new scheduled task to execute this command: 'C:\
      Program Files\Internet Explorer\IEXPLORE.EXE' 'http://local-
      host:2795/CaprikonRI1/reminder.aspx'. Check the option to enable
      the scheduled task to run at specified time (see Figure 6.6).
(3)   Configure the 'Schedule' to run the command on certain day, time
      and intervals (see Figure 6.7).
(4)   Configure the 'Setting' to stop the task if it runs for 1 minute (see
      Figure 6.8). This is to close the window that pops out after the task is
      executed.

Managing user information in the system

Challenge: The system should only allow the access of registered and
authorised users. There are two types of registered and authorised
users in the system (i.e. the PKM and other users). The PKM needs
to have the access to certain functions, such as that for adding new
projects or new members, validating knowledge, deleting a knowledge

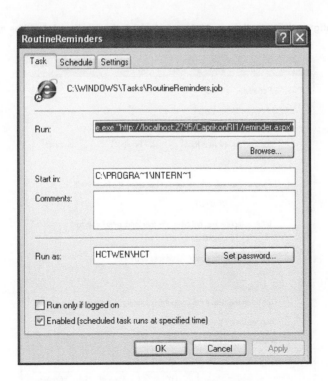

Figure 6.6   Configure the 'Scheduled Tasks' to execute the email script

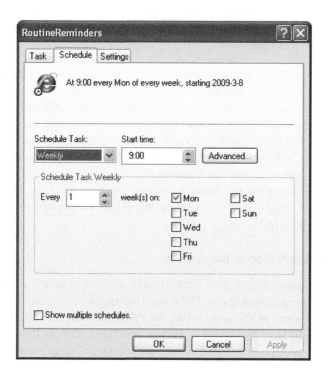

Figure 6.7   Configure the date, time and interval at which the email script will be executed

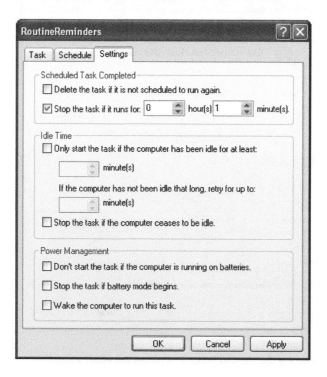

Figure 6.8   Configure the 'Scheduled Tasks' to close the pop out window after the task is executed

item from the database. These functions should not be made available to the other users.

Solution: Microsoft™ VWD Express 2005 provides a role-based authentication mechanism. This mechanism allows the customisation of certain parts of the system to make them available only to the users assigned with certain roles. All of the PKM will be assigned the 'Administrators' role, who will have access to the aforementioned functions (e.g. validating knowledge). Programme codes were written to distinguish the users with different roles when they attempt to access the restricted functions. For instance, before the function for validating knowledge is made available to the user, the associated programme will check to ensure that the user has an 'Administrators' role. The authentication mechanism can also prevent unauthorised or unauthenticated users from accessing the system (see Excerpt 6.7).

Dissemination of knowledge

Challenge: To facilitate 'live' sharing of knowledge captured, the system needs to disseminate the new knowledge captured through emails once the knowledge is entered. The emails should be sent only to the users who have opted for receiving the email notifications.

Solution: A data field was first created in a data table to store user's preference on whether they would like to receive email notifications. Programme codes were then written to send email notification to the users who have opted to receive email notifications when a knowledge item is added into the system. The programme codes will also attach a copy of the knowledge submitted together with its topic to the recipients in the email (see Excerpt 6.8).

```
Protected Sub  Page_Load(ByVal  sender  As Object,  ByVal  e  As  System.
EventArgs)  Handles Me.Load
    If User.IsInRole("Administrators") Then
        FormView4.Visible = True
        FormView4.DefaultMode = FormViewMode.Edit
    Else
        FormView4.Visible = False
    End If
End Sub
```

Excerpt 6.7   Programme codes to ensure that only PKM can access the knowledge validation function

```
Dim conn As New SqlConnection(ConfigurationManager.
ConnectionStrings
"ConnectionString").ConnectionString)
Dim sqlquery As String = "SELECT email from MemberInfo Where
([inMailingList] = 1)"
Dim cmd As New SqlCommand(sqlquery, conn)
Dim topic As String = topicTextBox.Text
Dim details As String = detailsTextBox.Text
Dim find As String = "(?<url>http://(?:[\w-]+\.)+[\w-]+(?:/[\w-
./?%&=]*)?)"
Dim Result As String = Regex.Replace(details, find,
"<a href=""${url}"">${url}</a>")
Dim objreader As SqlDataReader
Dim myVar As String

conn.Open()
  objreader = cmd.ExecuteReader(System.Data.CommandBehavior
.CloseConnection)
  myVar = ""
While objreader.Read()
  myVar += objreader("email") & ","
End While
  myVar = myVar.Substring(0, (myVar.Length - 1))

Try
  Dim mail As New MailMessage()
  mail.From = New MailAddress("h.c.tan@lboro.ac.uk")
  mail.To.Add(myVar)
  mail.Subject = "New Knowledge Item Added: " & topic
  mail.Body = Result & "<font face=Arial> You can access the
details of the knowledge through this link: <a href=http:
www.caprikon.org.uk>CAPRINET</a> </font>"
  mail.IsBodyHtml = True
  Dim smtp As New SmtpClient("ispstaff-mailout.lboro.ac.uk")
  smtp.Credentials = New NetworkCredential("user name",
"password")
  smtp.Send(mail)
  Catch ex As Exception
     Trace.Warn(ex.Message)
  End Try
```

Excerpt 6.8   Programme codes for sending email notification instantly when a new knowledge item is added

6.2.3 Database design

The Web-based knowledge base comprises two Microsoft™ SQL Server 2005 Express databases (i.e. the membership database and the main database).

The membership database contains the information about the membership, identity and authentication of users. It plays an important role in the security of the Web-based knowledge base. It helps to ensure that only the user with the correct user name, password and authorisation can access the stipulated sections of the knowledge base.

The main database stores all the details pertaining to reusable project knowledge. In the database, the details of a knowledge item are divided into a number of tables where each of the tables stores only one topic of information (see Figure 6.9). The data stored in the tables are linked by relations. This type of database structure (i.e. a normalised relational database) ensures that a non-primary key data is only stored in one table in a database. This helps to eliminate the potential of data update and deletion anomalies. Data update and deletion anomalies may happen if similar data is stored in two tables but the programme code only updates or deletes the data in one table. Details of the tables are as follows:

- Project details are stored in 'ProjectDetailsTable'.
- Details of reusable project knowledge are stored in 'Knowledge DetailsTable'.
- Details of the different categories of reusable project knowledge are stored in 'KnowledgeCategoryTable'.

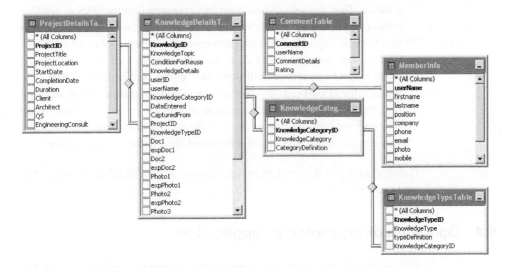

Figure 6.9   The main database's schema

- Details of the different types of reusable project knowledge are stored in 'KnowledgeTypeTable'.
- Comments for each of the reusable project knowledge are stored in 'CommentTable'.
- The users' personal contact details and preference are stored in 'MemberInfo' table.

6.3   Refinement of the IWS and user interface

The IWS and mock-up user interfaces were presented and reviewed by the case study companies in a mini-workshop conducted in a CAPRIKON Project Meeting. The user interfaces and the IWS were subsequently refined based on the findings of the mini-workshop. The prototype application was then developed based on the mock-up user interfaces and IWS created. The main outcomes of the mini-workshop include:

The idea of capturing reusable project knowledge from project meetings/reviews, and individuals were accepted. One of the workshop participants offered to use one of his/her company's projects to test the idea of capturing knowledge from the routine project meetings.

- The idea of capturing the rationale for making changes to documents was seen as crucial. However, it was not implemented in this version of the prototype. This is because it requires the full integration of the software prototype with an existing project extranet (i.e. one used for managing project documents), which was not immediately feasible.
- It was suggested that the validation of knowledge captured in the Web-based knowledge base could be conducted in the project meetings/reviews. This suggestion had been incorporated into the IWS as another option for knowledge validation. In addition, only the rating-based option would be incorporated into the prototype tool as the proof of concept.
- Companies were concerned about the function of the Web-based knowledge base which will send out email notifications when new knowledge is entered into the system. Regarding this, it was decided that the Web-based knowledge base should allow the users to choose whether they would like to receive the email notifications or not. This was also incorporated into the IWS and the prototype application.

The operation of the prototype application is described in detail in the next section.

6.4   Operation of the prototype application

This section describes the operation of the prototype application with the aid of relevant screen shots.

6.4.1   *Logging in*

When the prototype is started, the Login Page is displayed (see Figure 6.10). All the hyperlinks found on that page (except the 'Forgot Password?' link) will not function before the identity of the user is verified. The users can log into the system by entering their user name and password. In case they forget their password, they can click on the 'Forgot Password?' link. This brings up the 'Forgot Password Page' where the user will be requested to provide their login name and the answer to a secret question (see Figure 6.11). The password will then be sent to the user's registered email address in the system.

6.4.2   *Browsing the Summary Page*

After successfully logging into the system, the user will be redirected to the 'Summary Page' (see Figure 6.12). On the 'Summary Page', a list of the latest additions of knowledge items in the system and a list of knowledge items pending validation (i.e. either tagged as 'Draft' or 'To Be Revised') are shown. The knowledge topic, date on which it was entered, title of project from which it was captured, its current status (i.e. 'Draft' or 'To Be Revised') and an abstract of the knowledge are provided. If the user would like to know more about a listed knowledge item, the user

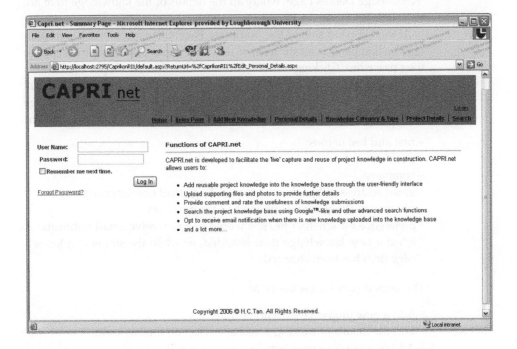

Figure 6.10   Login Page

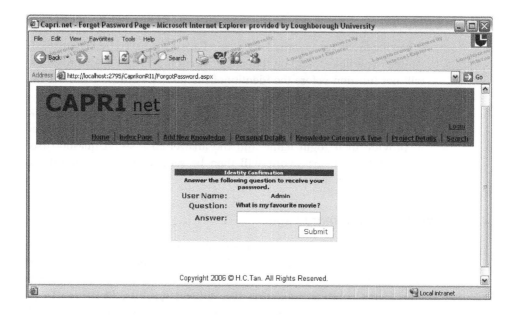

Figure 6.11   Forgot Password Page

can click on the 'read more' hyperlink. This will lead the user to the 'Knowledge Details Page' where all the details of the knowledge item are revealed (see Figures 6.19 and Figure 6.20).

At the left-hand side of the 'Summary Page' there are two coloured panels. The first panel is for the users to edit their personal details and to change their password. Clicking on either the link for editing one's personal details or the link for changing password will take them to the 'Edit Personal Details Page' (see Figure 6.13). In addition to allowing the user to change his/her password, the user can also edit the following personal information:

– first and last names
– position
– company
– email address (which is used for sending email notifications)
– phone (landline), mobile phone and fax numbers
– preference for whether he/she would like to receive email notifications when a new knowledge item is added, or when the status of a knowledge item has been changed.

The second panel is for the PKM to:

– Add a new project (see Section 6.4.5);
– Edit the details of existing project (see Section 6.4.5 as well);
– Add new users or members (see Section 6.4.7);
– Edit the details of the users.

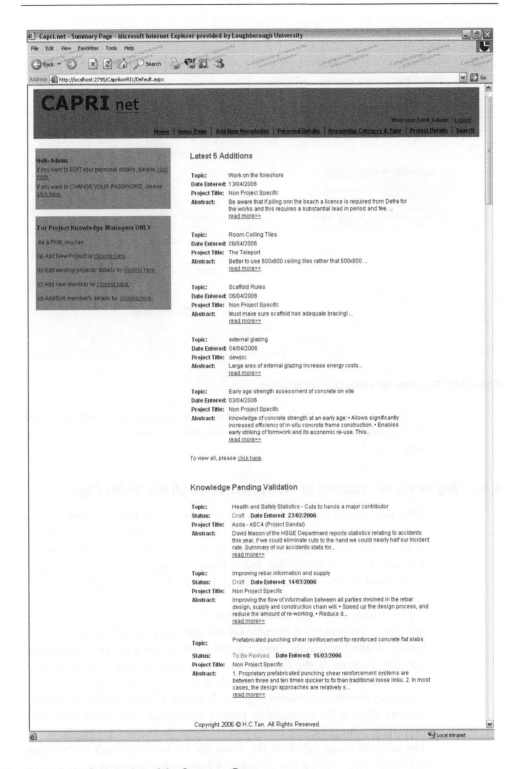

Figure 6.12   Screen shot of the Summary Page

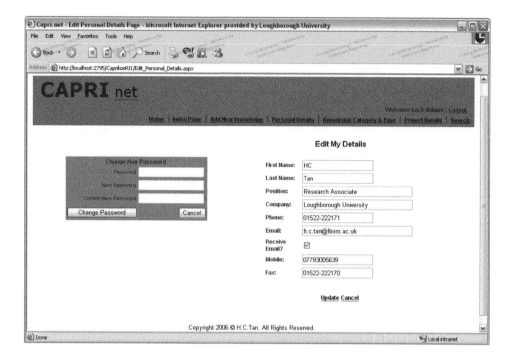

Figure 6.13   Template for editing personal details

If users other than the PKM attempt to access the functions, they will be informed that the functions are only accessible to the PKM (see Figure 6.14).

6.4.3   Exploring the content of the system through the 'Index Page'

If the user would like to have a complete view of the list of the knowledge captured in the system, he/she can click on the 'Index Page' link on top of any page. This redirects the user to the 'Index Page' (see Figure 6.15). The user can also access the 'Index Page' through the hyperlink located at the end of the list of the latest five knowledge additions on the 'Summary Page'.

The 'Index Page' comprises a shortcut menu to a list of knowledge items that fall under a particular knowledge category, and an index table listing all the knowledge items captured in the system. The knowledge category buttons of the shortcut menu are automatically created once a new knowledge category is created in the system. If the user clicks on a button on the shortcut menu (e.g. the 'Legal and Statutory Requirements' button), then a list of reusable project knowledge that belongs to that knowledge category will be shown in the index table (see Figure 6.16). The user can always click on the 'View All' button to go back to the original 'Index Page' with a complete list of knowledge captured.

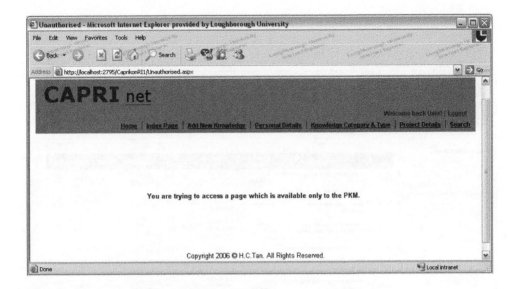

Figure 6.14   Screen shot of the Unauthorised Page

The details of knowledge items (i.e. topic, knowledge category and type, date entered, project title, status, author of the knowledge and the author's email address) are shown in the index table. On the 'Index Page', the user can:

– Click on the heading on the index table to change the way the knowledge captured are sorted (i.e. from ascending to descending order, and vice versa);
– Click on the 'ID' of a knowledge item in order to view the details of that knowledge (see Section 6.4.4 for details);
– Click on the 'Project Title' of a knowledge item to view the details about the project from which the knowledge is captured (see Figure 6.17);
– Click on the 'Author' to view the contact details of the author (see Figure 6.18). The information provided allows the user to contact the author of a knowledge item for further details;
– Click on the 'Click Here' email link to send an email to the author of the knowledge.

6.4.4 *Exploring and validating the details of a knowledge item*

The details of a knowledge item is revealed on the 'Knowledge Details Page' (see Figures 6.19 and 6.20) when the user clicks on the 'ID' of that knowledge on the 'Index Page' or other pages. Full details about the knowledge (including relevant image files) and the project from which

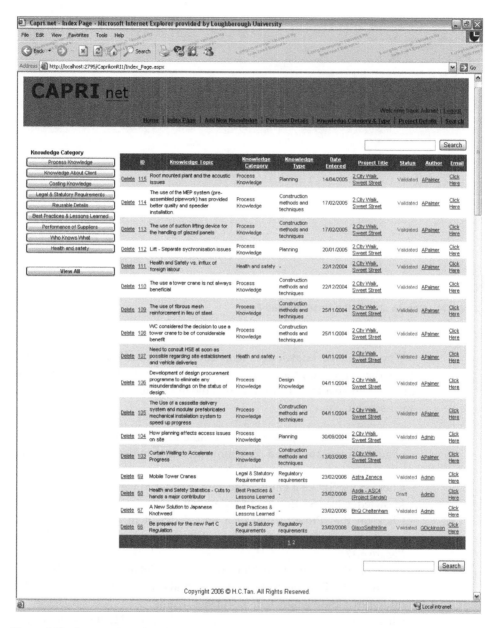

Figure 6.15   Screen shot of the 'Index Page'

the knowledge is captured are displayed on the page by default. There are also hyperlinks that can lead the user to the contact details of the author and other relevant websites. By clicking on the download links, the user can download and view the relevant documents. If the default size of the images displayed on the page is too small, the user can also click on the

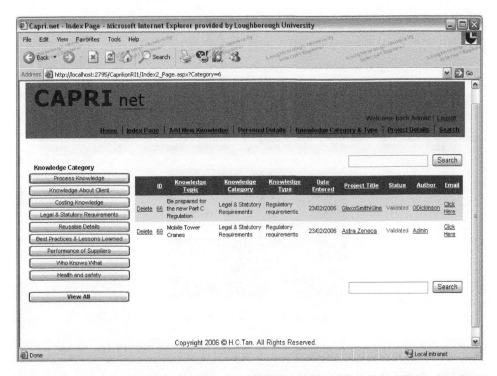

Figure 6.16   Screen shot of the Index Page showing list of knowledge about legal and statutory requirements

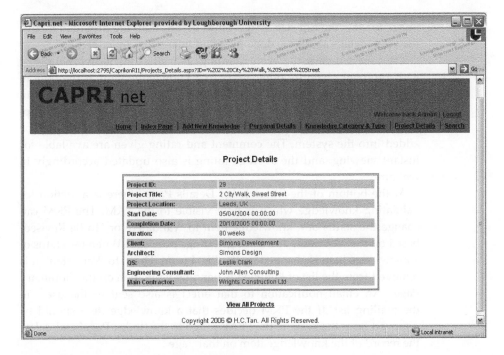

Figure 6.17   Screen shot of a project's details

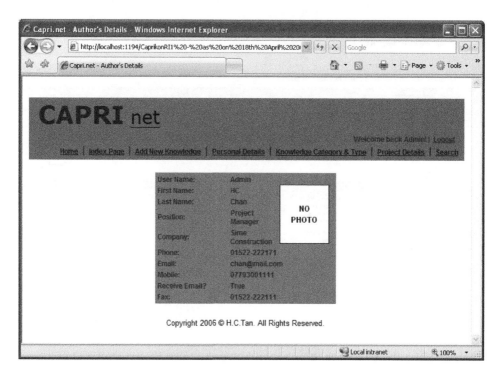

Figure 6.18   Screen shot of the page showing the author's contact details

image to view the full-size image. Short descriptions are provided to give the users details about the document and image files uploaded.

The user can view the average rating, as well as the individual rating and comment given by other users to a knowledge item at the bottom of the 'Knowledge Details Page'. The user can add his/her comment in the textbox provided, and select the rating to be given from the drop-down list. By clicking the 'Insert' button, the comment and rating will be added into the system. The comment and rating given are available for instant viewing, and the average rating is also updated accordingly in real time.

At the bottom of the 'Knowledge Details Page', there is a section for validating knowledge which is only visible to the PKM. The PKM can change the status of a knowledge item to 'Validated' or 'To Be Revised' based on the comments and average ratings provided. When the status of a knowledge item is changed by the PKM from 'Draft' to 'Validated', it is removed from the list of knowledge pending validation on the 'Summary Page'. An email notification to that effect is also sent to the users in the mailing list. If the PKM decides that a knowledge item should be removed from the system, the PKM can go to the 'Index Page' and delete the record of the knowledge item on that page.

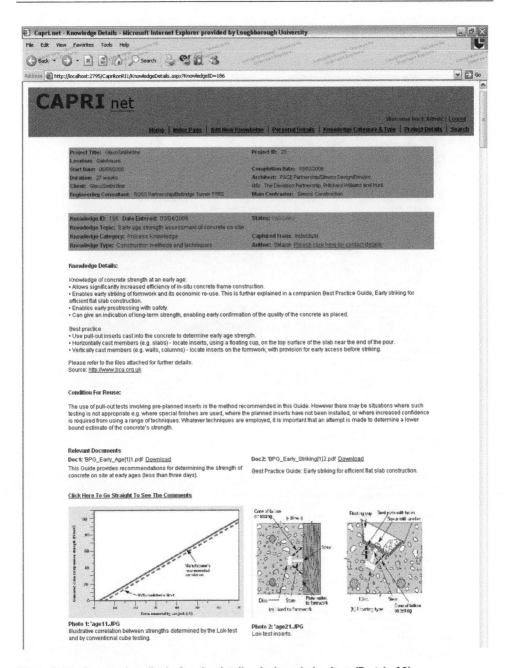

Figure 6.19   Screen shot displaying the details of a knowledge item (Part 1 of 2)

6.4.5   Add and Edit project details

The PKM can add a new project into the system by clicking on the 'add new project' link on the 'Summary Page'. The PKM will need to type in all the information into the textboxes provided, except the duration of

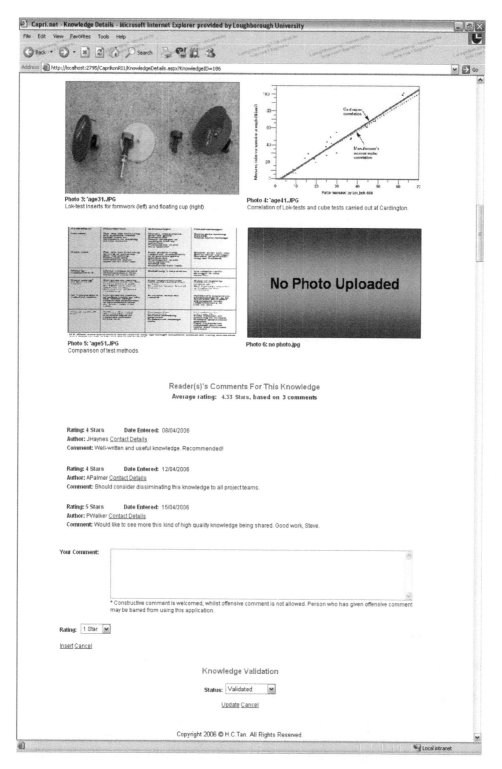

Figure 6.20 Screen shot displaying the details of a knowledge item (Part 2 of 2)

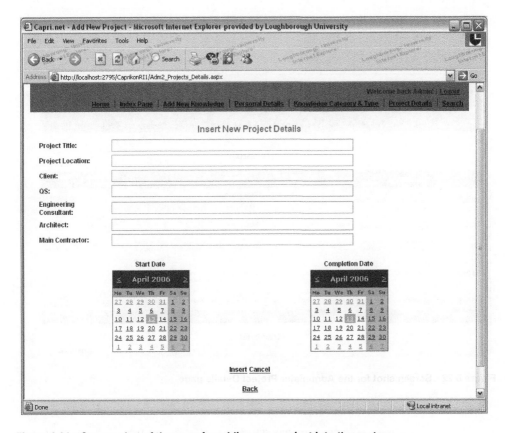

Figure 6.21   Screen shot of the page for adding new project into the system

the project which will be calculated automatically by the system from the start and completion dates entered (see Figure 6.21).

Occasionally, the PKM may need to edit or update the details of a project (e.g. changes in the start or completion date). In this case, the PKM can click to access the 'Administer Project Details Page' through the 'Project Details' hyperlink on top of every page. On that page, the PKM can click on the 'Edit' button which is only visible to PKM for editing the details of a project (see Figure 6.22).

6.4.6   Adding new knowledge category and type

Only the PKM is allowed to add new knowledge categories and knowledge types into the system. To access this feature, the PKM needs to click on the 'Knowledge Category & Type' link at the top of every page. This link redirects the PKM to the page for adding a new knowledge category

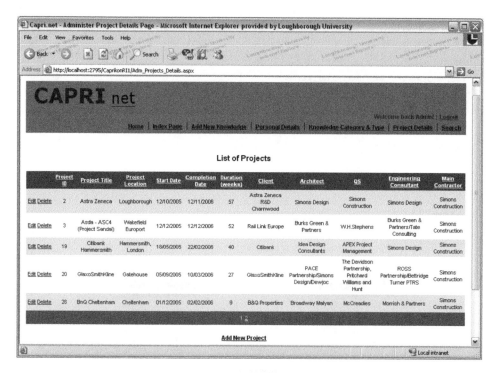

Figure 6.22   Screen shot for the Administer Project Details page

and type (see Figure 6.23). The procedure for adding a new knowledge category and type is as follows:

- The PKM enters a new knowledge category name and the associated definition. After the 'insert' button is clicked, the new knowledge category is created in the system.
- The PKM provides a name and definition for the new knowledge type. The PKM must select a knowledge category which the knowledge type belongs to from the dropdown list. The PKM can now click on the 'insert' button to add the new knowledge type into the system.

The details of the knowledge categories and types available in the system are revealed on the table at the top of the page.

6.4.7   Create account for new user

For security reasons, ordinary users are not allowed to create accounts for themselves in the system. Only the PKM has the authority to add new users. This function is accessible through the PKM-only section on the

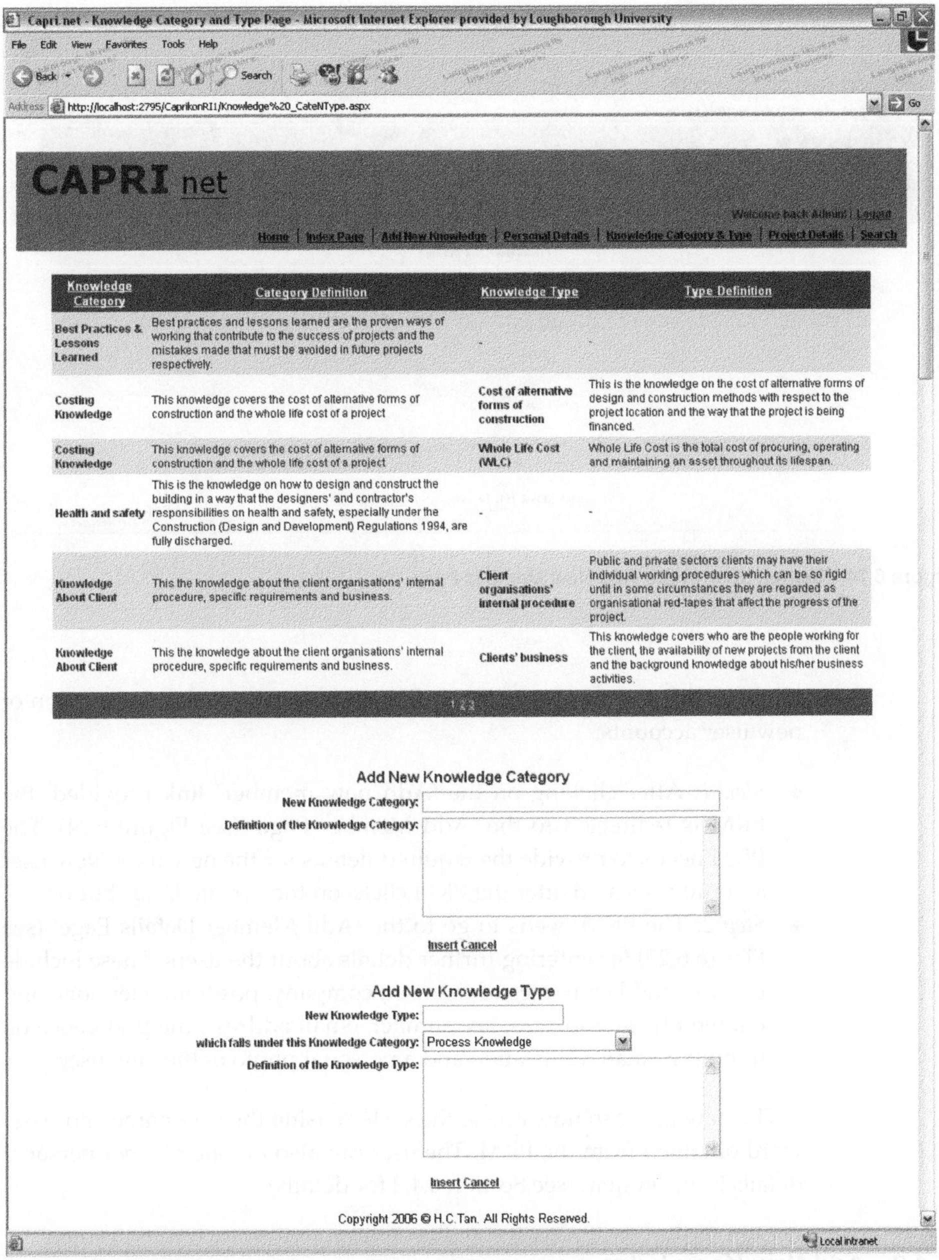

Figure 6.23   Screen shot for the Add New Knowledge Category and Type Page

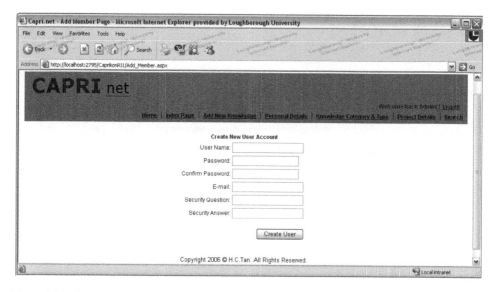

Figure 6.24   Screen shot of the Add New Member Page

'Summary Page' (see Figure 6.12). There are two steps for the creation of new user accounts:

- *Step 1*: After clicking on the 'Add new member' link provided, the PKM is redirected to the 'Add Member Page' (see Figure 6.24). The PKM needs to provide the required details for the new user. New user account is created after the PKM clicks on the 'Create User' button.
- *Step 2*: The PKM needs to go to the 'Add Member Details Page' (see Figure 6.25) for entering further details about the users. These include the first and last names of the user, company, position, telephone and mobile phone numbers, fax number, email address, the preference on receiving email notification and a personal photo of the new user.

The new user can now access the system using the user name and password obtained from the PKM. The user can also update his/her personal details in the system (see Section 6.4.2 for details).

6.4.8   Add New Knowledge

The 'Add New Knowledge Page' is accessible through the hyperlink available at the top of every page (see Figure 6.26). The page is characterised by three dropdown menus which provide a user-friendly means for entering repetitive information about a knowledge item (i.e. the project details, knowledge category and knowledge type). When a particular project is selected from the project title dropdown menu, the details of

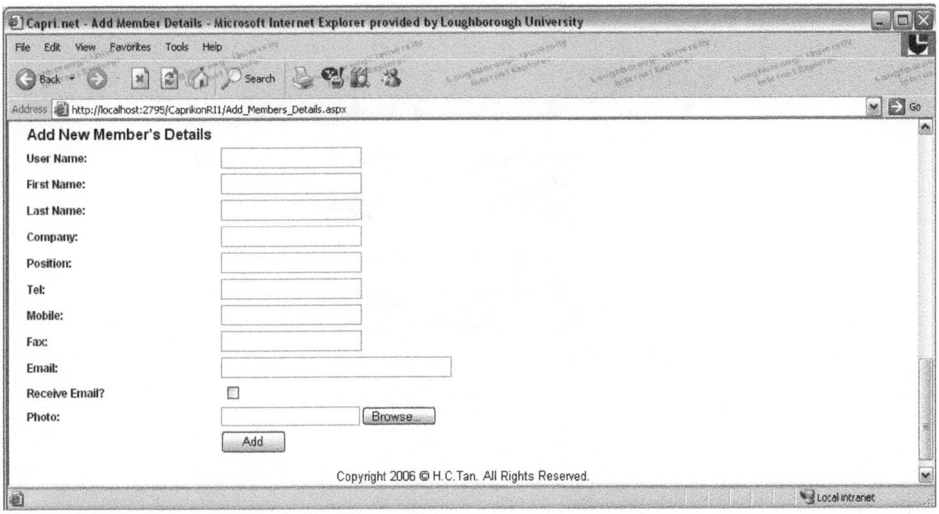

Figure 6.25   Screen shot for adding further details of a new member

the project are displayed at the bottom of the menu. This helps the users to ensure that they are referring to the right project (i.e. the one where the knowledge is captured from).

The dropdown menus for selecting the knowledge category and type are interconnected. When a particular knowledge category is selected by the user, the definition of the knowledge category is displayed. Meanwhile, a list of knowledge types that belong to that knowledge category will be shown in the dropdown menu for knowledge type. The definition for the selected knowledge type is also displayed.

There are two required field information that the user must complete (i.e. the knowledge topic and knowledge details) before clicking the upload button. Otherwise, a warning message will be shown (see Figure 6.27). Occasionally, there is some restriction as to the usage of a knowledge captured. For instance, a knowledge item which is captured from a Private Finance Initiative (PFI) hospital project may be only valid in the context of the PFI hospital projects. Therefore, the user should stipulate the restriction or condition for reusing a knowledge item in the section provided. If no condition for reuse is specified, this section will be tagged as 'Not Available' by the system. The user also needs to specify where the knowledge is captured from, that is either from project meetings/reviews or personal experience (individual).

Two upload functions are provided: one for uploading non-image files and another for image files. The user can click to select relevant images and other non-image files to upload into the system using the appropriate upload functions. The user can also provide a short description on the content of the document or image files in the textbox provided. If the user

Figure 6.26   Screen shot of the 'Add Knowledge Page'

is in doubt about what information should be provided in the various sections, a screen tip will pop out when the mouse cursor hovers over the respective yellow 'info' tag. After providing all the details, the user can click on the upload button to enter the knowledge into the system. An email notification will be sent instantly to all users who have opted to receive it.

6.4.9   Conducting a search

A user can search the Project Knowledge File using the Google™-like search function available at the top and bottom of the 'Index Page' (see Figures 6.15 and 6.28), or through the 'Advanced Search' function found on the 'Search Page' (see Figure 6.29). For the Google™-like search

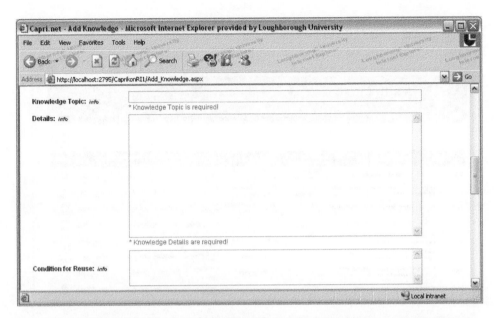

Figure 6.27   Screen shot of warning message if the required field information is not provided

function, the system will return all the results that contain the searched keyword(s). See Figure 6.28 for the Result Page of the Google™-like search function when the keyword 'concrete' is searched. Note that the 'delete' button in Figure 6.28 is visible only to the PKM.

If the user would like to limit the search results to the most relevant items only, then the advanced search function should be used. If the user selects the search button without specifying any search criteria, the search results returned will contain all the knowledge items captured in the system. If the user selects a particular project from the dropdown list before clicking 'search', the system will return a list of knowledge that was captured from that project only. However, the advanced search function also allows the user to narrow the search result by specifying a combination of details about the knowledge searched. These include:

– From which project the knowledge is captured; and/or
– The category and/or the type which the knowledge belongs to; and/or
– The status of the knowledge; and/or
– The topic of the knowledge; and/or
– The keyword(s) found in the details of the knowledge.

See Figure 6.30 for the screenshot of the results returned by the advanced search function.

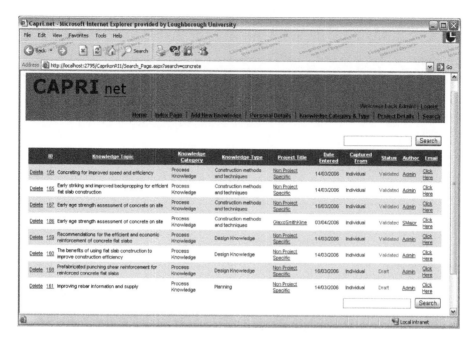

Figure 6.28   Screenshot of the Result Page of the Google™-like search function

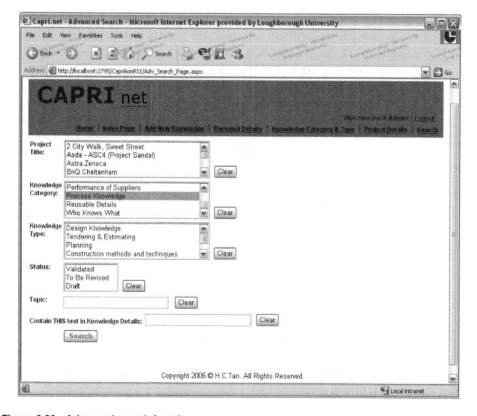

Figure 6.29   Advanced search function

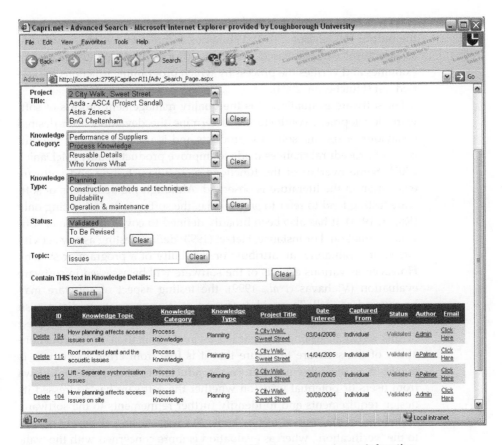

Figure 6.30   Screen shot of the results returned by the advanced search function

6.5   Testing and evaluation of Capri.net

Software testing is the process of executing computer software in order to determine whether the results it produces are correct (Glass, 1979) and to uncover evidence of defects (McGregor and Sykes, 2001). It is the examination of the behaviour of a program on sample data sets (Adrion *et al.*, 1982). According to Ould and Unwin (1986), two terms that are frequently used in the testing literature are *validation* and *verification*. Verification is the testing of an object against its specifications (Ould and Unwin, 1986), or 'Are we building the product right?' (Sommerville, 2001). Validation is the process of confirming that a deliverable matches the user's expectations (Ould and Unwin, 1986), and is concerned with 'Are we building the right product?' (Sommerville, 2001). An important classification of the tests available is the black-box and white-box dichotomy (Roper, 1994). Black-box techniques are also called 'functional' or 'specification-based' techniques (Roper, 1994). Black-box testing verifies whether the output is correct for a given input without verifying the process that produced the output (Hutcheson, 2003). White-box techniques may be referred to as 'structural' or 'code-based' techniques (Roper, 1994). White-box testing examines and verifies the process by which programme functions are carried out (Hutcheson, 2003).

For software evaluation, it is the quality management process of software development conducted to determine the deviation from desired behaviour of specific software products and is used to monitor the outcome of procedural changes made to improve product quality (McDaniel, 2002). Some overlap of the functions of software testing and software evaluation in the literature is observed. Although the definitions of software testing tend to refer to *program* (i.e. the software codes) testing only (Roper, 1994), it has also been broadly defined to cover the scope of software evaluation. For instance, Hetzel (1993) defines testing as any activity aimed at 'evaluating' an attribute or capability of a program or system. However, as various aspects of the software will be assessed in software evaluation (Vlahavas *et al.*, 1999), the testing aspect of software may be covered as well. To avoid confusion, it is appropriate to distinguish between the testing and evaluation of software. Hence, for the purpose of this research, testing is regarded as an examination of the functionalities of the software to ensure that it is free from error. Evaluation is regarded as the subsequent process conducted to obtain external views from users or potential users on whether the software has addressed its design requirements and to identify further refinements to the software. Therefore, in the context of this book, the term testing is closely related to the 'verification', whereas evaluation is more concerned with the 'validation' aspect. The details of the testing and evaluation undertaken are described in the following sections.

6.5.1   *Prototype testing*

The selection of tests to be performed is dependant on the aspects of the software to be tested (such as integration test to examine the communication between modules), and is restricted by time and resource constraints. Due to time constraints, extensive tests such as life-cycle testing (Roper, 1994) and hierarchical approach testing (McGregor and Sykes, 2001) which involve a series of different tests at different development levels of a software, were considered inappropriate. Moreover, as the main objective for the test was to ensure that the prototype will work as intended for the purpose of the evaluation, a Statement Test and ELHs Test (which examine the functions of the prototype against those stipulated in the design requirements) are more relevant. A Statement Test is conducted to ensure that all the sub-tasks attributed to a function work in the way they are supposed to do. ELHs Tests are conducted to ensure that all of the potential combinations of functions performed to the entity (e.g. reusable project knowledge) can be executed and will give the required result. In other words, the ELHs Test is to ensure that the whole system will work as expected. This combination of tests inclined towards the black-box approach. The details of the test procedures are described in the following sections.

Statement Test

This is the acceptance tests on the requirements (i.e. statements) for the design of software. Each of the requirements comprises an input action to be performed on the software application and an expected output of the input. In order to pass the test, the real output must match with the expected output. The details of the requirements of the prototype application that were tested are depicted in Table 6.2.

Table 6.2   Test results of the Statement/Requirement Test

Input	Expected output	Result
Log into the system using individual user name and password	• Login succeeded	Achieved
Log out from the system	• Log out succeeded	Achieved
Add project details into the system (PKM only)	• Project details added into system • Duration of project calculated automatically based on the start and completion dates entered	Achieved Achieved
Edit project details (PKM only)	• Project details edited • Duration of project updated automatically based on the start and completion dates entered	Achieved Achieved
Add new user (PKM only)	• New user added	Achieved
Add personal details into the system	• Personal details added into system	Achieved
Edit personal details	• Personal details edited	Achieved
Add a new knowledge category and its definition	• Knowledge category and its definition added into system	Achieved
Add a new knowledge type and its definition	• Knowledge type and its definition added into system	Achieved
Add a knowledge item, where the knowledge should be tagged as 'draft' or 'validated' based on its source	• Knowledge details added into system • Knowledge captured from individuals and groups are tagged as 'draft' knowledge and 'validated' knowledge respectively • Date of entering knowledge inserted automatically • Details of the author captured automatically • Index page and summary page updated • Email notification sent	Achieved Achieved Achieved Achieved Achieved Achieved
Delete a knowledge item (PKM only)	• Knowledge removed from system • Index page and summary page updated	Achieved Achieved

(Continued)

Table 6.2 (*Continued*)

Input	Expected output	Result
Search for a knowledge entry using:		
Google™-like search function	• Details of relevant knowledge, and the associated author and project details returned	Achieved
Advanced search function (i.e. through a combination of knowledge title, knowledge category, project name, and project type)	• Details of relevant knowledge, and the associated author and project details returned	Achieved
Access to knowledge		
Click to access the summary page (i.e. the default.aspx page)	• The summary and abstract of the latest five knowledge items added displayed	Achieved
	• List of the knowledge items that are either tagged as 'draft' or 'to be revised' shown	Achieved
Click to access a knowledge item	• Details of knowledge, and the associated author and project details returned	Achieved
Click to access relevant documents through the hyperlinks	• Dialogue box for either opening or downloading the files displayed	Achieved
Click on the relevant hyperlink to access the author's details	• Author's details displayed	Achieved
Click on the relevant hyperlink to send email to the author	• Depending on the default email client of the computer used, the template for writing email displayed	Achieved
Click on the photo/image shown on the page representing a knowledge item	• The photo/image displayed in its original (which is normally larger) size	Achieved
Click on the 'knowledge category' menu to access the knowledge that falls under a special category	• The list of all of the knowledge that fall under that category displayed	Achieved
Configure system on:		
Interval for sending routine email reminders (PKM only)	• Email reminders sent in accordance to the interval set	Achieved
The option for receiving email notifications when a new knowledge item is added	• Email notification sent to the user if he/she has chosen to receive email notification	Achieved
	• No email notification sent to the user if he/she has chosen not to receive email notification	Achieved
Knowledge validation		
Add comment about a knowledge item	• Comment added	Achieved
	• Details of the user who submitted the comment captured	Achieved
Add rating for a knowledge	• Rating added	Achieved
	• Average rating updated	Achieved
Access the knowledge validation function/control (PKM)	• The knowledge validation function/control is made visible to the PKM only	Achieved
Change the status of the knowledge from 'draft' to 'validated' (PKM only)	• Knowledge status changed; email notification sent	Achieved
	• Index page and summary page updated	Achieved
Change the status of the knowledge from 'draft' to 'to be revised' (PKM only)	• Knowledge status changed	Achieved
	• Index page and summary page updated; email notification sent	Achieved

The test results reveal that all the test inputs delivered the expected outputs. This means that the prototype application passed the Statement Test. Table 6.2 provides details of the test results. The prototype application was subsequently subjected to another test (i.e. the ELHs Test) before its evaluation was conducted. The details are presented in the specific section on ELHs Test.

ELHs Test

ELHs describe the life cycle of a knowledge item, from its creation in the system through all the actions performed on it, to its removal from the system (Roper, 1994). The life history of a knowledge item starts when it is being captured into the system, where it may subsequently be validated and hence shared, or removed from the system if rejected. While the knowledge item is captured in the system, it may be searched for and referenced any number of times. When the knowledge has become obsolete or is rejected, the knowledge item may be removed from the system.

The possible life histories experienced by a knowledge entity include:

- Possibility 1: Add, Search;
- Possibility 2: Add, Invalidate, Remove;
- Possibility 3: Add, Search, Invalidate, Remove;
- Possibility 4: Add, Validate, Search, Remove (when a knowledge item becomes obsolete);
- Possibility 5: Add, Search, Validate, Search, Remove (when a knowledge item becomes obsolete).

These generate the test requirements shown in Table 6.3 where the double horizontal lines separate distinct sets of test data. The test results are also shown in Table 6.3. The test result revealed that the prototype application passed the test on all the four possibilities of a knowledge item's life history. This shows that the prototype application can perform all the operations on a knowledge item as designed and that the prototype application is ready for the evaluation.

6.5.2 Prototype evaluation

This section describes the evaluation and associated results of the prototype application developed. The most useful features of the prototype application identified and the participants' suggestions for improvements are also presented.

Evaluation procedure

It is crucial to ensure consistency in the end-users' requirements identified from the case studies for the development of the methodology for 'live'

Table 6.3 **System test based on a test of the ELH (with and without built-in validation mechanism)**

System test		
Input	**Expected output**	**Result**
Possibility 1:		
Add a knowledge	Knowledge added into system; email notification sent; index page updated	Achieved
Remove the knowledge	Knowledge removed from system; index page updated	
Possibility 2:		
Add a knowledge	Knowledge added into system; email notification sent; index page updated	Achieved
Search for the knowledge	Knowledge details returned	
Remove the knowledge	Knowledge removed from the system; index page updated	
Possibility 3:		
Add a knowledge	Knowledge added into system; email notification sent; index page updated	Achieved
Validate the knowledge	Status of knowledge updated; index page updated; email notification sent	
Search for the knowledge	Knowledge details returned	
Remove the knowledge	Knowledge removed from the system; index page updated	
Possibility 4:		
Add a knowledge	Knowledge added into system; email notification sent; index page updated	Achieved
Search for the knowledge	Knowledge details returned	
Validate the knowledge	Status of knowledge updated; index page updated; email notification sent	
Search for the knowledge	Knowledge details returned	
Remove the knowledge	Knowledge removed from the system; index page updated	

capture of project knowledge and the prototype application which was developed accordingly for the purpose of the evaluation. To achieve this, the evaluation involved ten participants from four companies which were involved in the previous case study (i.e. Companies A, B, D and E). Of the ten participants, nine had participated in the case studies conducted which led to the development of the methodology. However, the tenth participant was also well informed of the development progress of the methodology through his/her colleague. The participants from the other two case study companies (i.e. Companies C and F) were unable to participate due to unforeseen circumstances.

Most of the evaluations were conducted on one-to-one basis with the exception of Company A. The evaluation started with a brief introduction of the concept of 'live' capture and reuse of project knowledge in construction. This was followed by the demonstration of the various features and operations of the prototype application. Subsequently, the participants were allowed to experiment with the prototype application by themselves. Guidance was given to the participants whenever necessary. An evaluation questionnaire was then given to the participants to complete. The details of the questionnaire are presented in the next section.

Questionnaire design

A questionnaire was designed to evaluate the prototype application against the end-users' requirements for developing the methodology for 'live' capture and reuse of project knowledge. The questionnaire comprised three main sections. Section A covered the background information about the participant while Section B consisted of 12 questions about the prototype application. The questions were further grouped into three subsections: Section 1 – Capture of knowledge, Section 2 – Representation of knowledge, Section 3 – Sharing/Reuse of knowledge and Section 4 – Ease of use. The participants were requested to provide their answers to the questions using a rating scale from 1 (very poor) to 5 (excellent). Section C provided an opportunity for the participants to identify the most useful features of the prototype application and to put forward their suggestions for improvements to the prototype application.

Evaluation results

The prototype application scored an average 3.9 out of 5.0 in the evaluation. The results, based on the analysis of the completed questionnaires, are depicted in Table 6.4. The average ratings of the various sections are presented in subsequent sections.

- *Capture of knowledge*
  The participants found the prototype application very capable of enabling the 'live' capture of project knowledge. A high average rating of 4.3 was given to this question. The participants were also highly satisfied with the prototype application's capability in terms of capturing the details of reusable project knowledge. This was evident by the average rating of 4.1 given by the participants. The third question (i.e. Question 1.3 in Table 6.4) on the validation mechanism of the prototype application received an average rating of 3.7. This reveals that the participants were confident that the adopted mechanism can help to ensure the accuracy and correctness of the reusable project knowledge captured. A satisfactory average rating of 3.4 was received on how well the overall methodology copes with avoiding the creation of additional workload. The comparatively lower average rating received was due to a rating of 2.0 given by two participants. They saw the need to adapt their companies' existing procedures to fully implement the methodology as an additional

Table 6.4   Ratings of key features of the prototype application

Sections	Average rating (out of 5)
Section 1: Capture of Knowledge	
1.1 How well does the system enable project knowledge to be captured 'live' (i.e. as soon as possible after a knowledge is created or identified)?	4.3
1.2 How complete is the system in capturing the details of a knowledge?	4.1
1.3 How well does the validation mechanism ensure the accuracy and correctness of knowledge captured?	3.7
1.4 How well does the system cope with avoiding the creation of additional workload?	3.4
Section 2: Representation of Knowledge	
2.1 How well does the index page provide users an overall view of all the knowledge captured in the system?	4.1
2.2 How well is knowledge organised in the system?	4.1
2.3 How well does the template represent the knowledge captured?	3.9
Section 3: Sharing/Reuse of Knowledge	
3.1 How well does the system facilitate the sharing of the project knowledge captured (i.e. through the provision of access to the knowledge captured through Web, links to additional information in the template for representing knowledge, the search function provided, etc.)?	4.1
3.2 How well does system achieve the concept of the 'live' sharing of project knowledge captured from a project (i.e. through allowing users to access of the knowledge via Web and sending email to users once new knowledge is added into the system)?	3.9
3.3 How reusable is the knowledge captured (i.e. during a subsequent project stage or on another project)?	3.6
Section 4: Ease of Use	
4.1 How good is the search function in locating the knowledge required?	4.1
4.2 How easy is the system to use overall?	4.3

workload. However, four of the ten participants acknowledged that it was impossible to totally avoid the creation of additional workload, but that created by the methodology was acceptable to them. This group of participants gave a rating of 3.0 for this question. Furthermore, there were also four participants who felt that the additional workload created was negligible. They gave a rating of 5.0, 4.5, 4.0 and 4.0 respectively.

● *Representation of knowledge*

The prototype application was recognised as very effective in representing the reusable project knowledge captured, where an average rating of 3.9 was given by the participants. Related to this, the participants also gave an average rating of 4.1 for both the way the Index Page provides users an overall view of all the knowledge captured in the system and the way knowledge is organised in the system.

- *Sharing/reuse of knowledge*

The methodology developed excelled in facilitating and realising the concept of 'live' sharing of the reusable project knowledge captured. It received an average rating of 4.1 and 3.9 for these areas respectively. The methodology was also found useful in facilitating the reuse of the project knowledge captured. This was evident by the average rating of 3.6 given by the participants to this question.

- *Ease of use*

The search functions (i.e. the Google™-like and the advanced search functions) received a high average rating of 4.1. This showed that the search functions of the prototype application are very efficient in helping the users to locate the required knowledge in the system. The prototype application was also perceived by the participants as very easy to use with a very high average rating of 4.3.

Suggestions for improvement

Participants described the prototype application as: 'overall excellent method of capturing knowledge', 'like the simplicity and general index page that offers quick links to knowledge', 'good interface design', 'good search function', 'format is clear', 'easy to use', 'not difficult to pick up and appears easy to work with' and 'easy to navigate'. They further pointed out that it is very easy to add relevant information and documents into the system, and that the 'time to add the knowledge is modest and will encourage use'. This proved that the methodology had successfully addressed the critical end-user requirements that significant additional workload is not desired. However, some suggestions for improvement were also received. These include:

(a) Create a mechanism for recording the quantity of hits on each knowledge item;
(b) Integrate built-in viewers for certain document types into the prototype application (e.g. Voloview for viewing AutoCAD files);
(c) Provide an audit trail for revisions made to the knowledge captured;
(d) Needs a disclaimer to indemnify the author of a knowledge item against the legal consequences of misuse or others relying on the accuracy and correctness of the knowledge.

A suggestion to automatically send out email notification when the status of a knowledge item is updated had been incorporated into the prototype application. The topic and textual details of the knowledge item can also be attached to the email notification sent. Suggestions (a), (b) and (c) entail extensive development in order to deliver the desired features and were hence not incorporated into this version of the prototype. The preparation of a disclaimer might be essential for knowledge capture activities involving different organisations, but this is better drafted by legal professionals.

Some suggestions were related to the future development of the prototype application. An evaluator suggested integrating the prototype application with other existing knowledge-based systems in an organisation. The prototype application will then become the core of the integrated system which allows a search across different systems to be conducted through it. Others saw the potential to commercialise the prototype application and proposed that a business case for this purpose be developed. These suggestions have been carefully considered and would be further explored in the future as appropriate.

7 Concluding Notes

This chapter concludes this book on capture and reuse of project knowledge in construction. It summarises the shortcomings of the current practices for managing project knowledge, briefly highlights the importance of a methodology for the 'live' capture and reuse of project knowledge and outlines the main user requirements and the development of the methodology. It concludes with salient points that researchers and practitioners need to take cognisance of, and identifies further work that can be conducted to enhance the methodology presented in the book.

7.1 Summary

This book covers the development of a methodology for 'live' capture and reuse of project knowledge in construction. The methodology reflects both the organisational and human dimensions of knowledge capture and reuse, and exploits the benefits of technology. The rationale for conducting the research on which the book is based was to address the knowledge loss problem due to the time lapse in capturing important knowledge from a project. The aim was achieved following the successful development and the positive evaluation result received for the aforementioned methodology. The methodology developed comprises a Web-based knowledge base, and an Integrated Workflow System (IWS) and a Project Knowledge Manager (PKM). Various methods were used to achieve the objectives of the research, namely extensive literature review, case study, workshop and the evaluation of the methodology developed. The results achieved through these methods are summarised below.

Extensive literature review on knowledge management (KM) was first carried out to gain the essential understanding on the subject in sufficient detail. KM processes which comprise knowledge capture, knowledge sharing, knowledge reuse and knowledge maintenance were proposed. The literature review also revealed the heavy reliance on post project reviews, the reassignment of people across projects and the contractual and organisational arrangements for the transfer of knowledge in the construction industry. The various shortcomings of the existing practice in managing project knowledge effectively were uncovered. Capturing knowledge through post project reviews was found less successful mainly due to the time constraints for conducting it upon the completion of a project and the knowledge loss due to time lapse in capturing the knowledge gained. Reassignment of people from one project to another was undermined by the high staff turnover in the industry and the weakness of human memory in memorising facts which it depends on for knowledge transfer. The attempt to facilitate the sharing of knowledge through contractual and organisational arrangements also suffered from the fact that organisations collaborating on one project might be in competition in another project which makes them reluctant to share critical knowledge. This was further aggravated by corporate restrictions on the transmission of knowledge and information.

Further review of existing literature suggested the need for a methodology that facilitates the capture and reuse of important knowledge from ongoing projects once it is created or identified (i.e. 'live') across geographical boundaries. It was established that the methodology could address the aforementioned knowledge loss problem which is due to time lapse in capturing knowledge from projects and staff turnover. Furthermore, it would enable the knowledge captured from the initial stages of a project to be reused at the later stages of the project, and help to seize every knowledge reuse opportunity which would, in turn, help to maximise the value of reusing the knowledge captured.

The novelty of the aforementioned methodology was confirmed in subsequent literature reviews. A number of research projects conducted to address the various issues of managing knowledge in construction were identified from existing literature. However, these research projects were focused at strategic and business perspectives, specific types of knowledge, specific project phases or specific types of construction organisation. The need for an approach which is capable of capturing project knowledge, irrespective of the type of project, the type of construction organisation and project phases and particularly capturing the knowledge 'live', has not been adequately addressed. Research at Stanford (Reiner and Fruchter, 2000) was considered as being closest to the goal of 'live' capture and reuse of project knowledge. Nonetheless, the research did not cover the entire project but focused only on the design evolution stage. Hence, a case for developing the methodology was made.

Various concepts were explored, in particular learning histories (Kleiner and Roth, 1997) and Collaborative Learning (Digenti, 1999), in terms of the capability to facilitate the 'live' capture and reuse of project knowledge. The study of Collaborative Learning had indirectly led to the design of the methodology to capture knowledge through project meetings or reviews. This was because Collaborative Learning showed that the interactions among the members of a team can help to reveal and share their tacit knowledge. An insight was also gained from 'learning histories' on how a standard format for representing the knowledge, as required by the case study companies, can possibly be designed.

The nature and characteristics of reusable project knowledge, the current practice for capturing reusable project knowledge in construction and the end-user requirements for the design of the methodology were identified through the six case studies conducted. A wide spectrum of reusable project knowledge was identified from the case study. The knowledge identified were aligned and grouped into the following categories: Process Knowledge, Knowledge of Clients, Costing Knowledge, Knowledge of Legal and Statutory Requirements, Knowledge of Reusable Details, Knowledge of Best Practices and Lessons Learned, Knowledge of Performance of Suppliers, Knowledge of Who Knows What and Other Types of Knowledge. Reusable project knowledge was found to exist as a mix of tacit and explicit knowledge, rather than as distinctive tacit or explicit knowledge alone. As a result, the methodology was designed to explicate tacit knowledge into explicit knowledge as far as possible and to help to connect people to the very tacit knowledge which is extremely difficult to explicate.

The end-user requirements for the development of the methodology for 'live' capture and reuse of project knowledge in construction identified were as follows:

- The methodology must facilitate the capture and access of project knowledge 'live' (i.e. as soon as possible once knowledge is created or identified) and across geographically dispersed offices.
- The methodology should not create significant additional cost and workload to the companies.
- An appropriate legal framework is required to overcome the client's potential restriction or copyright problem on the sharing of knowledge.
- A validation mechanism is required to ensure the accuracy and correctness of knowledge before it is shared.
- A standard format for representing the knowledge which contains the background information on the project, abstract, details, conditions for reuse and reference is required.

Various KM techniques and technologies used by the case study companies for the capture of reusable project knowledge were investigated. The KM techniques used were post project reviews, communities of practice (CoPs), reassignment of people, research collaboration, partnership-like arrangements, preparation of reusable details, research and development, team meetings, road shows, presentations, workshops, succession management and mentoring. The KM technologies used were Groupware, custom-designed software, expert directory and project extranet. The KM techniques and technologies used have their strengths and shortcomings, and in fact complement each other. Hence, a combination of KM techniques and technologies was selected as the most viable option for meeting the requirements for the development of a 'live' knowledge capture and reuse methodology.

A Web-based knowledge base was found to be the closest to meeting the end-user requirements identified. The reasons are as follows:

- A Web-based knowledge base allows knowledge to be entered and accessed 'live'.
- No significant additional cost is required due to the pervasive use of intranets and the capability of the Web-based knowledge base to run on the existing intranet/Internet systems.
- The Web-based knowledge base can be designed to be as user-friendly as possible and to capture some information automatically in order to avoid the creation of significant additional workload.
- A mechanism can be built into the knowledge base for monitoring the validation of knowledge submitted as a means of ensuring its accuracy.
- Standard templates can be created to ensure that project knowledge is entered in accordance with the format specified.
- It can provide the necessary platform for accessing and sharing knowledge which is captured in the form of video clips and other multimedia formats. It may be used in conjunction with other Web-based applications (e.g. Groupware and video conferencing tools) to enhance the sharing of knowledge, particularly the tacit knowledge.

Project meetings and reviews were chosen to capture knowledge in a group setting to ensure that a more holistic and more complete set of knowledge is captured. For security reasons and the fact that the prototype application would be accessed by users from different organisations, the prototype application was designed to run in the extranet environment.

The methodology developed, which comprises a Web-based knowledge base, an IWS and a PKM as the administrator, allows project knowledge to be captured 'live' from ongoing projects. The methodology was encapsulated into a prototype application using ASP.NET Visual Basic 2.0 and Microsoft™ SQL Server 2005 Express Edition combination. The development of the prototype application was influenced by the Web IS Development Methodology (WISDM), which is specifically devised for the development of Web-based Information Systems including Web-based knowledge base. A mini-workshop was conducted to refine the IWS and the design of the user interface prior to the commencement of the programming tasks. This helped avoid the potential introduction of changes to the design of the prototype application at later stages which might lead to significant reworks and delays. The prototype application was demonstrated to the ten participants from four organisations in the evaluation to assess the extent to which it had met the end-user requirements identified from the case study. A high overall average rating of 3.9 out of 5.0 together with positive comments from the participants were received in the evaluation. The practicality of the prototype application was also indirectly confirmed by the suggestions to commercialise the prototype application given by the participants. This signified the achievement of the research's aim.

7.2 Conclusions

Based on the findings in the book, the following conclusions are drawn:

(1) The importance of KM has been recognised by the construction industry with various KM techniques and technologies adopted for the capture and reuse of knowledge learned from projects. However, the industry still faces serious knowledge loss mainly due to:
   o The time lapse in capturing important knowledge from projects;
   o High staff turnover in the industry;
   o The lack of an organised and systematic approach to the capture and sharing of reusable knowledge from projects.

(2) Information and communication technology (ICT), particularly Web-based technology, is crucial in realising the concept of 'live' capture and reuse of project knowledge. The capture of reusable project knowledge is often confronted by the lack of a methodology to allow individuals to upload their knowledge at anytime which is convenient to them, and to access the knowledge captured at any place when they need it. The Web-based prototype application addressed these limitations and enables reusable project knowledge to be captured 'live' from an ongoing project at anytime and any place, and to be shared in real time across geographical offices.

(3) The methodology for 'live' capture and reuse of project knowledge in construction developed in this study can help to minimise the knowledge loss problem by capturing project knowledge in the most-timely way before the details are forgotten or the project team disbanded. By making project knowledge available for reuse once it is captured in the system, it helps to seize every knowledge reuse opportunity. Hence, it also helps to maximise the value of reusing the knowledge captured through 'live' reuse.

(4) Reusable project knowledge in construction often exists as a mix of tacit and explicit knowledge, rather than as distinctive tacit or explicit knowledge alone. Any methodology developed for managing reusable project knowledge must therefore be capable of facilitating the capture and reuse of both tacit and explicit dimensions of

the knowledge. Generally, KM technologies are better for managing explicit knowledge, whereas KM techniques are crucial for the sharing of tacit knowledge. However, it was noticed that KM technologies and techniques are complementary, and best deployed in concert. For instance, a KM technology such as 'skills directory' can help to connect the people with tacit knowledge with those who need the knowledge. On the other hand, a KM technique such as 'communities of practice' can help to direct the project team members to the right knowledge base (i.e. a KM technology) where the explicit knowledge is stored. Hence, the synergy of both KM technology and technique is required in order to better manage either tacit or explicit knowledge. Therefore, the methodology developed attempts to capture the collective view of the knowledge learned and facilitate the sharing of tacit dimension of the knowledge from a project through project meetings and reviews (i.e. a KM technique), and utilises a Web-based knowledge base (i.e. a KM technology) to facilitate the 'live' capture and reuse of the knowledge.

(5) Cost and workload are the main concerns of construction organisations for the implementation of a KM system. As it is notoriously difficult to justify the return on investment (ROI) for the implementation of a KM system, most organisations favour a cautious approach and require that no significant additional cost and workload are created for the purpose.

(6) It was observed that the implementation of KM in the construction industry is very often executed without a detailed strategy or a clear understanding of what exactly needs to be done in order to achieve the aims. Consequently, most of the approaches were adopted in a piece-meal manner without an overall strategy for how the various approaches can be synergised. This leads to the unnecessary waste of resources and less satisfactory results. It is crucial to have an overall strategy supported by the top management and a detailed action plan in order to address this problem.

7.3 Limitations of the research

There are some limitations of the methodology and prototype system for live capture and reuse of project knowledge in construction. These include:

(1) The prototype application which was developed using Microsoft's ASP.NET 2.0 can only run on Microsoft's Window's platform. Although some third party developers have developed some programmes to enable the applications developed using ASP.NET 2.0 to run on non-Windows platforms (e.g. Mono developed by Novell), these solutions are not mature yet.

(2) It is recognised that wider validation of the prototype system is needed as it only involved the same sources of data collection (i.e. the case study companies). In addition, there was also a lack of full representation from contractors' organisations as the views of contractors were mainly obtained from the construction division of a case study company and other interviewees based on their previous working experience in contractor organisations.

(3) It is also recognised that more time is required to fully evaluate the prototype application. This will allow more projects to be used in the evaluation to help improve the richness of the contents captured. Furthermore, this will also enable the users to provide more constructive suggestions for improvements through personal experience of using it over a reasonably long period of time. However, this was not possible in this project due to time constraints.

7.4 Further work

For the prototype application to be effectively used in a commercial environment, a number of additional features would need to be incorporated into the system whilst some existing features would need to be improved. These include:

- Improvement of the existing search functions in the prototype application. Although this received a very high rating in the evaluation conducted, the search function can be further improved to support Boolean queries (e.g. user can search for 'hospital' AND 'PFI' NOT 'London'). The search function may even be extended to incorporate technology such as text-mining. This will help increase the precision of the search function.
- Improvement of the membership control to restrict the viewing of knowledge captured to those projects in which the companies are involved. This is an advanced feature not available in the prototype application.
- Development of the function for the capture of the rationale for making changes to documents as outlined in the IWS of the methodology. This will entail the full integration of the prototype application with an electronic document management system (or a project extranet).
- Development of the other knowledge validation options (e.g. comment-based option and majority opinion-based option). These options should be made available to the PKM when configuring the system for the capture of knowledge.
- Development of a mechanism for recording the number of hits on each knowledge item.
- Integration of built-in viewers for certain document types into the prototype application (e.g. Voloview for viewing AutoCAD files).

- Development of a mechanism to provide an audit trail for revisions made to the knowledge captured.

A number of areas for further research are identified. These include the following:

- Explore the integration of the prototype application with other existing information systems of an organisation (e.g. customer relationship database, internal human resource database which contains staff details and other technical information databases). This may involve the development of a middleware application to streamline and automatically link the relevant information in other information systems with that in the prototype application. This can help to synergise the benefits brought about by the different information systems and to further enhance the richness of the prototype application's contents.
- Investigation to fully automate the methodology for 'live' capture of project knowledge. This may involve the use of software agents, which are capable of learning from the patterns of usage (e.g. through analysing the most searched keywords in the prototype application) and understanding the relationships between terminologies, to automatically locate and disseminate relevant project knowledge from the Internet and intranet in real time. The software application developed can be complemented by a robust way of representing the results returned to help people to understand the contents.
- Investigation of methods for integrating the personal KM systems of project team members with the organisational KM systems. This may cover the development of a software application which allows individuals to capture their personal knowledge in a specified format, and share some of the knowledge (as desired) with others. This will allow others to tap into the knowledge of their colleagues in a way that is not possible through existing approaches. The software application developed may also have a built-in mechanism to assess the level of an individual's involvement and contribution to the organisational knowledge base, which can be linked to the staff appraisal system as appropriate.
- Development of a system for helping a project team to identify the relevant knowledge captured and to learn from the knowledge captured prior to the start of a new project. Currently, the explicated project knowledge is often captured in the printed form (e.g. best practice guide), and also in the form of electronic files (e.g. video and sound clips). There is a need to develop an effective methodology to represent the knowledge captured in various forms to improve the learning of project team members.
- In CoPs, best practice and lessons learned are shared, and new knowledge can be created through the discussions amongst the members. However, although the knowledge captured or shared within a

community of practice may also be important to the members of other CoPs, the existence of the knowledge is very often only available to its members. As a result, the opportunities for reusing the knowledge are not fully seized and the potential resultant benefits of reusing the knowledge are not fully exploited. Therefore, it is critical for construction organisations to know what its CoPs know, and also to create a mechanism to facilitate the sharing of relevant knowledge from all CoPs prior to commencing new projects. This will allow the collective knowledge of relevant lessons learned and best practices from all the CoPs to be shared to avoid the repetition of similar mistakes, and to better manage new projects right from the beginning.

Further exploration of the commercial potential of the prototype application is also recommended.

7.5   Concluding remarks

The importance of KM in construction, in particular the 'live' capture and reuse of project knowledge, was evident through this research. The nature and characteristics of reusable project knowledge, the shortcomings of current practice in managing project knowledge and end-user requirements as to the development for a methodology for 'live' capture and reuse of project knowledge in construction, were identified from the case studies conducted. The findings led to the development of the aforementioned methodology for 'live' capture and reuse of project knowledge, which was subsequently encapsulated in a Web-based prototype application. The application of the methodology developed can help to minimise the knowledge loss problem whilst enabling the knowledge captured to be reused widely to maximise the benefits to construction project teams.

Appendix A

Table Comparing the Various Knowledge Management Process Models

Author(s)	Knowledge management processes
Robinson *et al.* (2001)	• Discovering, locating and capturing • Organisation and storage • Sharing and transferring • Modifying and applying • Archiving and retirement
Kululanga and McCaffer (2001)	• Acquiring • Creating • Sharing • Storing • Utilising
Rollett (2003)	• Planning • Creating • Integrating • Organising • Transferring • Maintaining • Assessing
Tiwana (2000)	• Acquisition • Sharing • Utilisation
Bhatt (2001)	• Creation • Validation • Presentation • Distribution • Application
Mertins *et al.* (2001)	• Create • Store • Distribute • Apply
Soliman and Spooner (2000)	• Create • Capture • Organise • Access • Use
Davenport and Prusak (2000)	• Knowledge generation: acquisition, dedicate resources, fusion, adaptation and knowledge networking • Knowledge codification and coordination • Knowledge transfer

Appendix B

Details of the Types of Reusable Project Knowledge Identified

Reusable project knowledge	Details of the knowledge	Current practice to capture the reusable project knowledge
• Process knowledge	**• Design** Design knowledge can be subdivided into two categories: generic design knowledge and specialist design knowledge. Generic design knowledge focuses on the use of standard approach in design to ensure that designers have taken into consideration the issues like health and safety, and designer's risk assessment as required by regulations into the design. The latter category, that is specialist design knowledge, is the expert knowledge required in the design of special facility such as pharmaceutical facility. This knowledge covers what the necessary systems and facilities are and how they work. The possession of this knowledge enables the designer to produce a complete design although some of the necessary facilities are not mentioned or are omitted in the client's brief. In addition, this knowledge may enable a company to come out with alternative design options for the client. **• Tendering and estimating** This knowledge covers the assignment of proposals/operations manager for the task, the making of decision to submit tender, preparation of estimate, establishing overall strategy and framework for bid, risk and opportunity analysis, tender adjudication (deciding the mark-up for the tender) and further negotiation. **• Planning** This knowledge is concerned with the sequence and duration of construction activities, as well as the estimated total time required to construct a particular design. This knowledge is also important in providing advice to the client on the impact of his/her decision to the duration of the project. The planning of a project can be based on the successful program of other similar type of projects. This can help to reduce the time required for planning as compared to starting from scratch.	(1) Generic design knowledge It is captured in designer's standard procedures to ensure that they have taken into consideration all criteria and issues while preparing the design. (2) Specialist design knowledge This knowledge remains tacit in the head of people and is reused through the reassignment of expert to other projects. In addition, this knowledge is also transferable through demonstrating to others how to do it. (1) This knowledge is shared through the estimators CoPs and informal discourse. (2) It is also captured in database system containing the prices of each of the elements. (1) This knowledge is captured through personal experience and feedback from contractors and subcontractors. (2) This knowledge is also partly captured in the planning application software used.

● **Construction methods and techniques**

This comprises the knowledge of:

(a) Various construction methods or techniques available and the suitability of these methods to a project. This also covers the cost, speed, requirements in terms of human resources and technology, as well as the constraints imposed by the project's individual characteristics on the method of construction.

(b) Previous mistakes made on the selection of construction methods which are to be avoided.

(c) Influence of material selection to the construction of the facility. The decision made by designers on the selection of materials, such as the type of roof, affects the methods, process and speed of construction.

(d) Other factors that have impacts on construction.

(1) This knowledge is captured through hands on experience and often remain tacit. It is normally reused by reassigning people to other projects.

(2) It is also captured through careful selection of staff to ensure that only those who have good understanding on construction process are recruited.

● **Buildability**

This is the knowledge on the 'optimum integration of construction knowledge and experience in planning, design, procurement and field operation to achieve overall project objective'. It is important to capture this knowledge as there is evidence that construction quality and productivity on site are affected by the buildability of design.

This knowledge is captured in the head of people.

● **Operation and maintenance**

The growing importance of Private Finance Initiative (PFI) projects has significantly contributed to the need to capture the knowledge on managing the operation and maintenance of the facility built. This knowledge also covers the customers' feedbacks regarding the delivery of the building and their experience on occupying the building, the 'learning on which part of the design works and otherwise', as well as the influence of the design on the maintenance and operation of the facility.

This knowledge is captured in the post project review (PPR).

(Continued)

163

Reusable project knowledge	Details of the knowledge	Current practice to capture the reusable project knowledge
● **Knowledge about client**	(a) Clients' requirements There are two types of clients' requirements: general and specific. Clients from similar sector are very likely to have similar set of general requirements. For instance, the clients from education sector normally require that a rather similar set of class room facilities to be provided. However, due to difference in the nature of business, individual client also tends to have specific requirements or preference which are to be followed and incorporated into the design. This knowledge helps improve designer's capability to understand client's brief which in turn enables the designer to develop a design that better addresses the client's needs. (b) Client organisations' internal procedure Public and private sectors clients may have their individual working procedures which can be so rigid until in some circumstances they are regarded as organisational red-tapes that affect the progress of the project. Therefore, it is important to identify the red-tapes and the potential impacts in advance if interruption to the progress of project is to be avoided. (c) Clients' business This knowledge covers about who are the people working for the client, the availability of new projects from the client and the background knowledge about his/her business activities.	(1) Knowledge about client's standard procedures and operational constraints can be captured in procedural manual of the company. (2) Knowledge about clients specific requirements normally remain tacit and captured in people's head. This knowledge can be shared through formal meeting and discourse with the project team who has worked with the client before. (3) Other specific types of knowledge under this category can be captured in PPR.
● **Costing knowledge**	(a) Cost of alternative forms of construction This is the knowledge on the cost of alternative forms of design and construction methods with regard to the project location and the way that the project is being financed. (b) Whole Life Cost WLC is the total cost of procuring, operating and maintaining an asset throughout its lifespan. There is evidence that many public and private sectors clients now procure on WLC rather than capital cost. The ability of a company to prepare a design with a low WLC is dependant on the wealth, currency and accuracy of this knowledge.	(1) Costing knowledge can be captured in the estimating software and is also available through the subscription to the relevant website, for example, Building Cost Information Service (www.bcis.co.uk). (2) Knowledge on Whole Life Cost (WLC) is accessible by subscribing to the relevant web-based services such as the WLC Forum (www.wlcf.org.uk) or captured through internal cost information.

• Knowledge of legal and statutory requirements

(a) Regulatory requirements

This knowledge covers the requirements and responsibilities imposed by British Standards, Code of Practice, etc. on the clients, designers and contractors. The regulatory requirements change regularly over time. Thus, all the parties have to be aware of the changes and the impacts to their practice in order to abide by the new requirements.

(b) Health and safety

This is the knowledge on how to design and construct the building in a way that the designers' and contractor's responsibilities on health and safety, especially under the Construction (Design and Development) Regulations 1994, are fully discharged.

(c) Contract

The terms and conditions of contracts must be continuously evaluated to suit changes.

(1) This knowledge is available through subscription to the relevant web service and in the form of CD.

(2) It is also captured by sending representative to attend external course to understand the impacts of latest changes of regulations to current practice and then disseminate the learning within the company.

• Knowledge of reusable details

Reusable details consist of:

(a) Standard design details
(b) Specifications
(c) Method statements

Standard design details are such as the design drawings of specific areas and associated fittings of a facility. The reuse of the design details, specifications and method statements helps to avoid the reinvention of the wheel and also leads to time and cost savings. Adaptations might be necessary for the reuse of the details. Time saved can be used for making improvements to the details.

The chances to reuse the details are dependant on the proportion of similar type of projects, and the degree of repeating business from the same client.

(1) The standard reusable design details and specifications are captured in the drawings and specifications of a project respectively. The reuse of such details may require the person to contact the originator of the documents for further explanation on the rationale of design and the context for reuse.

(2) Some companies use a more formalised approach where sessions are held for the project team working under the same client to identify the areas where standard details on design and specification can be created. The standard details created are then made available to others in electronic form for reuse.

• Knowledge of best practices and lessons learned

Best practices and lessons learned are the proven ways of working that contribute to the success of projects and the mistakes made that must be avoided in future projects respectively. These are also referred to as the factors of 'success and failure' of project by one of the companies. Best practices and lessons learned are among the most common types of knowledge captured by construction organisations.

This knowledge is normally captured in the PPR and other reviews and meetings conducted at the end of the various project stages. The findings are then compiled as the company's best practice guide and code of practice.

(Continued)

Reusable project knowledge	Details of the knowledge	Current practice to capture the reusable project knowledge
• **Knowledge of performance of suppliers and KPIs**	**(a)** Performance of suppliers The suppliers referred to are other consultants, contractors, subcontractors, material suppliers, etc. that is anyone who has contributed services or goods to the project. The capture of this knowledge facilitates better selection of suppliers for future projects. **(b)** Key Performance Indicators (KPIs) KPIs are used to evaluate the performance of a project. The results of the evaluation can be used as a bench mark for continuous improvement in other projects.	This knowledge is captured by carrying out a qualitative assessment based on a predetermined set of criteria at the end of project. The result can be fed into a custom designed database and made accessible for reuse through intranet. This is captured through internal reviews, PPRs and collaboration with other companies.
• **Knowledge of who knows what**	This is the knowledge on the skills, experience and expertise of each of the members of staff. This knowledge is crucial as it impinges the successful reuse of other knowledge. It serves as a guide to lead people to the right source or right people with the knowledge. It assists in connecting people to people for the sharing of knowledge, particularly the tacit knowledge which is difficult to codify.	This knowledge is captured in the members of staff's personal file, curriculum vitae or personal web page in company's intranet.
• **Other types of knowledge**	• **Risk management** This is about the associated risk of working with a particular client and suppliers in a particular area with particular contractual arrangement and time constraint. • **Team working** This is the knowledge on how to manage a team and to prevent the relationship breakdown. This knowledge is more concerned with the project management rather than the specific issues or areas of the construction project. • **Project management** This is concerned with how to improve the performance of projects.	This knowledge is captured in the risk assessment report and the application software used to conduct risk assessment. Part of this knowledge is captured in the case study or history of the projects. People to people interactions play important role in the sharing of this knowledge. This knowledge is captured in people's head.

Appendix C

Additional Learning Situations Related to Change Management, Problem-Solving and Innovation

	Learning situations/triggers of knowledge production	Type
Peer (2002)	The most common reasons for change order, which result in changes, are: • Change in scope: for instance, client has requested a design change. • Unforeseen condition: for instance, the site conditions differ from expected. • Professional errors and omissions: for instance, the professional has incorrectly drawn the construction design plans and specifications.	Change management
Lazarus and Clifton (2001)	Sources of change: • Legislative change, for instance reduction in the acceptable discharge rates into external drainage. • Design change, for instance change to the cladding system which leads to further amendments in design. • Client change, for instance provision required by the client for further expansion of facility. • Contractor change, for instance contractor proposes different methods of construction for a section of work. • Site conditions change, for instance existing foundation design has to be revised due to unexpected ground conditions. There are two other factors which may lead to change: • Inflation or relative price rise; • Difficulties with contractors.	Change management
Park (2002)	Changes in work state; processes and methods that deviate from original construction plan or specifications.	Change management

(Continued)

	Learning situations/triggers of knowledge production	Type
Trauner (1993)	Trauner (1993) identifies a number of problem situations. However, only those directly relevant to learning situations are addressed here. The problem situations are: • Termination and default; • Projects behind schedule; • Claims and disputes; • Budgets-related issues such as over budget.	Problem-solving

McLoughlin _et al._ (2000)	McLoughlin _et al._ (2000) identify a range of economic drivers which organisations have to respond to when there is any change. These drivers can be grouped into change management and problem-solving related triggers of knowledge production.
	
	(1) Change management related triggers:
	
	    • Changing market requirements, such as demands for time compression and requirements for whole life project management.
	
	    • Regulation/de-regulation and environmental issues, that is the impacts of changes in regulatory requirements to the project.
	
	(2) Problem-solving related triggers:
	
	    • New sources of competition, particularly when moving into a market requiring new capabilities.
	
	    • Human resources issues, for instance the changes in the supply and cost of labour force from the local labour market.
	
	(3) Innovation related trigger:
	
	    • Fundamental and invasive technology improvements which have an effect on the economics of the project during the course of its lifetime.
		Change management, problem-solving and innovation

Egbu (2002)	Egbu (2002) identifies that those listed below are important types of innovations in project-based organisations: • New technology that has internal benefits to the company. • New process that has benefits to the company. • New approach to providing services to customers/clients. • New procedures for obtaining goods/services. • New product that provides competitive advantage for the company. • New external relations, for example partnering, joint ventures. • New administrative policy for example incentive schemes, bonuses.	Innovation

Appendix D

Companies' Practice and Requirements on Knowledge Representation

Company	How project knowledge is represented
A	(a) General headings are provided on the type of project knowledge.
	(b) Case studies or detailed explanation of the knowledge to help others to understand and hence reuse the knowledge.
	(c) The conditions for reusing the knowledge must be made clear to the users.
	(d) Checklists to show:
	• The issues relevant to the particular project.
	• The characteristics of the project that are related to the context for the reuse of the knowledge.
B	Sharing the bullet-point learning in a Web environment, each with a short description prepared to give the audience basic background information. This is supplemented by video clips to capture the detailed explanation from the originator of the learning.
C	Establishing convenient means, such as people's personal profile and knowledge network aided by custom-designed IT-systems, for people to communicate with each other and share their knowledge. Some knowledge of technical and contractual issues are represented in the form of 'feedback notes' in accordance with the format specified. The 'feedback notes' are made available to the members of staff over the company's intranet.
D	A standardized approach is required. The knowledge captured must be organised and represented in a logical and simple to understand way, and readily accessible to others within the organisation. Knowledge on know-how to perform a specific task (such as how to approach difficult situations) can be captured in the organisation's standard procedures.
E	The methodology developed for capturing or representing the knowledge should avoid the introduction of excessive additional workload to people. The additional workload created should be integrated into daily job functions and be carried out within normal working hours.
F	Knowledge represented comprises two sections:
	• Context of the knowledge such as the type of project and project stage, where the knowledge is concerned, and an explanation of how to reuse the knowledge.
	• The financial impact, such as the cost saving if the suggestion is implemented.
	Some process knowledge can be represented in the form of interactive process maps.

References

Aamodt, A. and Plaza, E. (1994) Case-based reasoning: Foundational issues, methodological variations, and system approaches. *Artificial Intelligence Communications*, 7(1), 39–59.

Adrion, W.R., Branstad, M.A. and Cherniavsky, J.C. (1982) Validation, verification, and testing of computer software. *ACM Computing Surveys*, 14(2), 159–192.

Al-Ghassani, A.M. (2002) *Literature Review on KM Tools*, Technical Report, Department of Civil and Building Engineering, Loughborough University, UK.

Allen, W., Bosch, O., Kilvington, M., Oliver, J. and Gilber, M. (2001) Benefits of collaborative learning for environmental management: Applying the integrated systems for knowledge management approach to support animal pest control. *Environmental Management*, 27(2), 215–223.

Anumba, C.J. and Evbuomwan, N.F.O. (1998) Collaborative working in construction – The need for effective Communication Protocols. *Proceedings 4th Computing Congress of the American Society of Civil Engineers*, Philadelphia, USA, 16–18 June, 1997, pp. 89–96.

Ardichvili, A. Page, V. and Wentling, T. (2003) Motivation and barriers to participation in virtual knowledge-sharing communities of practice. *Journal of Knowledge Management*, 7(1), 64–77.

Argyris, C. and Schon, A. (1978) *Organisational Learning*. Prentice-Hall, Englewood Cliffs, NJ.

Augenbroe, G., Verheij, H. and Schwarzmüller, G. (2002) Project Web sites with design management extensions. *Engineering, Construction and Architectural Management*, 9(3), 259–271.

Avison, D. and Fitzgerald, G. (2003) *Information Systems Development: Methodologies, Techniques and Tools*. McGraw-Hill, London.

Barson, R.J., Foster, G., Struck, T. *et al.* (2000) Inter- and intra-organisational barriers to sharing knowledge in the extended supply-chain. In: B. Stanford-Smith and Kidd, P.T. (eds), *E-business – Key Issues, Applications, Technologies*, pp. 367–373. IOS Press, Oxford.

Becerra-Fernandex, I. and Sabherwal, R. (2001) Organizational knowledge management: A contingency perspective. *Journal of Management Information Systems*, 18(1), 23–55.

Beijerse, R.P. (1999) Questions in knowledge management: Defining and conceptualizing a phenomenon. *Journal of Knowledge Management*, 3(2), 94–109.

Belecheanu, R., Pawar, K.S., Barson, R.J., Bredehorst, B. and Weber, F. (2003). The application of case based reasoning to decision support in new product development. *Integrated Manufacturing Systems*, 14(1), 36–45.

Bhatt, G., Gupta, J.N.D. and Kitchens, F. (2005) An exploratory study of groupware use in the knowledge management process. *The Journal of Enterprise Information Management*, 18(1), 28–46.

Bhatt, G.D. (2001) Knowledge management in organizations. *Journal of Knowledge Management*, 5(1), 68–75.

Bhatt, G.D. and Zaveri, J. (2002) The enabling role of decision support systems in organisational learning. *Decision Support Systems*, 32(3), 297–309.

B-Hive (2001) *Building a higher value construction environment*. (B-Hive Project Website at http://is.lse.ac.uk/b-hive/) [accessed 21/10/2004]

Billett, S. (2003) Workplace mentors: Demands and benefits. *Journal of Workplace Learning*, 15(3), 105–113.

Björkegren, C. (1999) Learning for the next Project: Bearers and barriers in knowledge transfer within an organisation, Linköping Studies in Science and Technology, Thesis No. 787, The International Graduate School of Management and Industrial Engineering, No. 32, Licentiate Thesis, Linköping, Sweden.

Blacker, F., Reed, M. and Whitaker, A. (1993) Editorial: Knowledge workers and contemporary organization. *Journal of Management Studies*, 30(6), 851–862.

Bollinger, A.S. and Smith, R.D. (2001) Managing organizational knowledge as a strategic asset. *Journal of Knowledge Management*, 5(1), 8–18.

Boyd, D., Egbu, C., Chinyio, E., Xiao, H. and Lee, C.C.T. (2004) Audio diary and debriefing for knowledge management in SMEs, *Proceedings of ARCOM 20th Annual Conference*, Heriot Watt University, UK, 1–3 September, pp.741–747.

Bresman, H., Birkinshaw, J. and Nobel, R. (1999) Knowledge transfer in international acquisitions. *Journal of International Business Studies*, 30(3), 439–463.

Bresnen, M. (1996) An organisational perspective on changing buyer–supplier relations: A critical review of the evidence. *Organisation Articles*, 3(1), 121–146.

Bresnen, M., Edelman, L., Newell, S., Scarbrouugh, H. and Swan, J. (2002) Knowledge management for project-based learning in construction, *10th International Symposium – Construction Innovation and Global Competitiveness*, W65/W55, September 9–13, Cincinnati, Ohio, USA, pp. 172–184.

Bresnen, M., Edelman, L., Scarbrough, H. and Swan, J. (2003) Social practices and the management of knowledge in project environments. *International Journal of Project Management*, 21(3), 157–166.

Bullinger, H.J., Worner, K. and Prieto, J. (1997) Wissensmanagement heute. Daten, Fakten, *Trends*, Fraunhofer-Institu fur Arbeitswirtchaft und Organisation (IAO), Stuttgart.

Burke, R.J., McKeen, C.A. and McKeenna, C. (1994) Benefits of mentoring in organizations: The mentor's perspective. *Journal of Managerial Psychology*, 9(3), 23–32.

Burrel, G. and Morgan, G. (1979) *Sociological Paradigms and Organizational Analysis*. Heinemann, London.

Burton-Jones, A. (1999) *Knowledge Capitalisation: Business, Work, and Learning in the New Economy*. Oxford University Press, Oxford.

Byham, W.C. (2002) HEADSTART: A new look at succession management. *IVEY Business Journal*, May/June, 10–12.

CAPRIKON Report (2004) *Knowledge capture and re-use in construction projects: Concepts, practices and tools*, In: P.M. Carrillo, J.M. Kamara, C.J. Anumba, N.M. Bouchlaghem, C.E. Udeaja, and H.C. Tan (eds), University of

Newcastle & Loughborough University, Research Report, 2004, ISBN: 1 897911 28 9.

Carrillo, P.M. (2004) Managing knowledge: Lessons from the oil and gas sector. *Construction Management and Economics*, 22(6), 631–642.

Carrillo, P.M. and Anumba, C.J. (2000) *Knowledge Management for Improved Business Performance*, Engineering and Physical Sciences Research Council (EPSRC) Grant Ref: GR/N01330.

Carrillo, P.M., Robinson, H.S., Anumba, C.J. and Al-Ghassani, A.M. (2003) IMPaKT: A framework for linking knowledge management to business performance. *Electronic Journal of Knowledge Management*, 1(1), 1–12.

Carrillo, P.M., Robinson, H.S., Al-Ghassani, A.M. and Anumba, C.J. (2004) Knowledge management in UK construction: Strategies, resources and barriers. *Project Management Journal*, 35(1), 46–56.

Chao, G.T., Walz, P.M. and Gardner, P. (1992) Formal and informal mentorships: A comparison on mentoring functions and contrast with nonmentored counterparts. *Personnel Psychology*, 45(3), 619–636.

Christensen, K.S. and Bang, H.K. (2003) Knowledge management in a project-oriented organization: Three perspectives. *Journal of Knowledge Management*, 7(3), 116–128.

CIPD (2004) *Recruitment, Retention and Turnover 2004: A Survey of the UK and Ireland*, Chartered Institute of Personnel and Development.

CIRIA (2004) *Business case for knowledge management: Guidance & toolkit for construction* (Project website at http://www.ciria.org/rp700.htm) [accessed 21/10/2004]

Cirovic, G. and Cekic, Z. (2002) Communications and forum: Case-based reasoning model applied as a decision support for construction projects, *Kybernetes*, 31(6), 896–908.

Clawson, J.G. and Kram, K.E. (1984) Managing cross-gender mentoring, *Business Horizons*, 27(3), 22–31.

Construction Best Practice (CBPP) (2004) *Introduction to knowledge management*. <www.cbpp.org.uk> [accessed 27/02/2004]

Cross, R. and Baird, L. (2000) Technology is not enough: improving performance by building organizational memory. *Sloan Business Review*, 41(2), 69–78.

Crossan, M.M., Lane H. and White, R. (1999) An organizational learning framework: From intuition to institution. *Academy of Management Review*, 24(3), 522–537.

C-SanD (2001) *Creating, sustaining and disseminating knowledge for sustainable construction: Tools, methods and architecture* (C-SanD Project Website at http://www.c-sand.org.uk) [accessed 20/10/2004]

Cuthell, J.P. (2005) Beyond collaborative learning: Communal construction of knowledge in an online environment, *Web Information Systems and Technology International Conference (WEBIST)*, Miami.

Davenport, T.H. (1998) *Some principles of knowledge management*. <http://www.mccombs.utexas.edu/kman/kmprin.htm> [accessed 21/10/2004]

Davenport, T.H. and Prusak, L. (2000) *Working Knowledge: How Organizations Manage What They Know*. Harvard Business School Press, Boston, Massachusetts.

Davenport, T.H., De Long, D.W. and Beers, M.C. (1997) *Building Successful Knowledge Management Projects*. Ernst & Young LLP.

De Long, D. (1997) Building the knowledge-based organization: How culture drives knowledge behaviors, *Working Paper*, Ernst & Young's Center for Business Innovation, Boston.

Demarest, M. (1997) Understanding knowledge management. *Long Range Planning*, 30(3), 374–384.

Dennis, A.R., Valacich, J.S., Connolly, T. and Aynne, B.E. (1996) Process structuring in electronic brainstorming. *Information Systems Research*, 7(2), 268–277.

Dent, R.J. and Montague, K.N. (2004) *Benchmarking Knowledge Management Practice in Construction*. CIRIA, London.

Diakoulakis, I.E., Georgopoulos, N.B., Koulouriotis, D.E. and Emiris, D.M. (2004) Towards a holistic knowledge management model. *Journal of Knowledge Management*, 8(1), 32–46.

Digenti, D. (1999) *The Collaborative Learning Guidebook*. Learning Mastery, Somerville, MA.

Disterer, G. (2002) Management of project knowledge and experiences. *Journal of Knowledge Management*, 6(5), 512–520.

Dixon, N.M. (2000) *Common Knowledge: How Companies Thrive by Sharing What They Know*. Harvard Business School Press, Boston, MA.

Drucker, P.F. (1993) *Post-Capitalist Society*. Butterworth-Heinemann, London.

Drucker, P.F. (1998) The coming of the new organization. In: *Harvard Business Review on Knowledge Management*, pp. 1–19. Harvard Business School Press, Boston.

Duffy, J. (1996) Collaborative computing, groupware and knowledge. *Information Management & Computer Security*, 4(2), 39–41.

Earl, M. (2001) Knowledge management strategies: Toward a taxonomy. *Journal of Management Information Systems*, 19(1), 215–233.

Ebbinghaus, H. (1885) *Uber das Gedachtnis*, Leipzig, Dunker. Translated by Ruger, H.A. and Bussenius, C.E. (1913), Teachers College, Columbia University. <http://psychclassics.yorku.ca/Ebbinghaus/index.htm> [accessed 16/08/2004]

Egan, J. (1998) *Rethinking Construction*, United Kingdom, pp. 1–39.

Egbu, C. and Botterill, K. (2001) Knowledge management and intellectual capital: Benefits for project based industries. In: J. Kelly and K. Hunter (eds), *Proceeding of the RICS Foundation and Building Research Conference (COBRA)*, School of the Built and Natural Environment, Glasgow Caledonian University, UK, 3–5 September, Vol. 2, pp. 414–422.

Egbu, C., Botterill, K. and Bates, M. (2001) The influence of knowledge management and intellectual capital on organisational innovations, *Proceeding of the 17th Annual Conference of the Association of Researchers in Construction Management (ARCOM)*, University of Salford, UK, 5–7 September, Vol. 2, pp. 547–556.

Egbu, C.O. (2002) Knowledge management, intellectual capital and innovation: Their association, benefits and challenges for construction organisations, *Proceedings of the 10th International Symposium on Construction Innovation and Global Competitiveness*, W65/55, Cincinnati, Ohio, USA, 9–13 September, Vol. 1, pp. 57–70.

Engström, T.E.J. (2003) Sharing knowledge through mentoring. *Performance Improvement*, 42(8), pp. 36–42.

Eppler, M.J. and Sukowski, O. (2000) Managing team knowledge: Core processes, tools and enabling factors. *European Management Journal*, 18(3), pp. 334–341.

Fiol, C. and Lyles, M. (1985) Organizational learning. *Academy of Management Review*, 10(4), 803–813. In: A. Jashapara (2004), *Knowledge Management: An Integrated Approach*, Financial Times Prentice Hall, UK.

Fiol, C.M. (1994) Consensus, diversity, and learning in organisations. *Organisational Science*, 5(3), 403–420.

Fong, P.S.W. (2003) Knowledge creation in multidisciplinary project teams: An empirical study of the processes and their dynamic interrelationships. *International Journal of Project Management*, 21, 479–486.

Fruchter, R., Reiner, K., Yen, S. and Retik, A. (2000) KISS: Knowledge and information slider system, *Proceedings of Construction Information Technology, CIT'2000*, Island, June.

Gallupe, R.B. (2001) Knowledge management systems: Surveying the landscape. *International Journal of Management Reviews*, 3(1), pp. 61–67.

Gill, T.G. (1995) High-tech hidebound: Case studies of information technologies that inhibited organisational learning. *Accounting, Management and Information Technology*, 5(1), pp. 41–60.

Glass, R.L. (1979) *Software Reliability Guidebook*. Prentice-Hall, Englewood Cliffs, NJ.

Goh, S.C. (2002) Managing effective knowledge transfer: An integrative framework and some practice implications. *Journal of Knowledge Management*, 6(1), pp. 23–30.

Gokhale, A. (1995) Collaborative learning enhances critical thinking. *Journal of Technology Education*, 7(1), Fall'95.

Goodman, R.E. and Chinowsky, P.S. (2000) Taxonomy of knowledge requirements for construction executives. *Journal of Management in Engineering*, 16(1), pp. 80–89.

Gotschall, M. (2000) E-learning strategies for executive education and corporate training. *Fortune*, 141(10), pp. S5–S59.

Green, S., Newcombe, R., Fernie, S. and Weller, S. (2004) *Learning Across Business Sectors: Knowledge Sharing between Aerospace and Construction*, BAE Systems, UK.

Gulliver, F.R. (1987) Post-project appraisals pay. *Harvard Business Review*, March–April, 128–132.

Hall, J., Sapsed, J. and Williams, K. (2000) Barriers and facilitators to knowledge capture and transfer in project-based firms, *Proceedings of 4th International Conference on Technology Policy and Innovation*, Curitiba, Brasil, 28–31 August. <http://in3.dem.ist.utl.pt/downloads/cur2000/papers/S28P04.PDF> [accessed 12/02/2003]

Hamada, T. and Scott, K. (2001) Anthropology and international education via the Internet: A collaborative learning model. *The Journal of Electronic Publishing (JEP)*, 6(1), University of Michigan Press. <http://www.press.umich.edu/jep/06-01/hamada.html> [accessed 27/02/2004]

Hamilton, A. (2001) *Management Projects for Success: A Trilogy*, First Edition, Thomas Telford, UK.

Hamilton, I. (2005) *Project collaborative extranets for construction: A guidance note*. <http://www.cica.org.uk/extranet-docs/cica-construction-extranets.pdf> [accessed 03/08/2005]

Hansen M.T. and Davenport, T.H. (1999) *Knowledge Management at Andersen Consulting*. Case No. 9-499-032, Harvard Business School, Boston.

Hansen, M.T., Nohria, N. and Tierney, T. (1999) What's your strategy for managing knowledge? *Harvard Business Review*, March/April, pp. 106–116.

Hari, S., Egbu, C. and Kumar, B. (2005) A knowledge capture awareness tool: An empirical study on small and medium enterprises in the construction industry. *Engineering, Construction and Architectural Management*, 12(6), 533–567.

Harman, C. and Brelade, S. (2000) *Knowledge Management and the Role of HR: Securing Competitive Advantage in the Knowledge Economy*. Financial Times Prentice Hall, London.

Harvey, M., Palmer, J. and Speier, C. (1998) Implementing intra-organisational learning: A phased-model approach supported by Intranet technology. *European Management Journal*, 16(3), pp. 341–354.

Hasan, H. and Gould, E. (2001) Support for the sense-making activity of managers. *Decision Support Systems*, 31, pp. 71–86.

Health and Safety Executive (HSE) (1997) *Construction (Design and Management) Regulations 1994: The health and safety file*, HSE Information Sheet; Construction Sheet No. 44. <http://www.hse.gov.uk/pubns/con44.htm> [accessed 11/10/2003]

Hearst, M. (1999) Untangling text data mining, *Proceedings of ACL'99: The 37th Annual Meeting of the Association for Computational Linguistics*, University of Maryland, USA, 20–26 June, pp. 3–10.

Hearst, M. (2003) *What is text mining?* <http://www.sims.berkeley.edu/~hearst/text-mining.html> [accessed 01/08/2005]

Heisig, P. and Vorbeck, J. (2001) Cultural change triggers best practice sharing – British Aerospace plc. In: K. Mertins, P. Heisig and J. Vorbeck (eds), *Knowledge Management: Best Practices in Europe*, Springer-Verlag, Berlin.

Heisig, P., Berg, C. and Drtina, P. (2001) Open minded corporate culture and management supports the sharing of external and internal knowledge phonak. In: K. Mertins, P. Heisig, and J. Vorbeck (eds.), *Knowledge Management: Best Practices in Europe*, Springer-Verlag, Berlin.

Hernández-leo, D., Villasclaras_Fernández, E.D., Asensio_Pérez, J.I. and Dimitriadis, Y. (2006) COLLAGE: A collaborative learning design editor based on patterns. *Educational Technology and Society*, 9(1), 58–71.

Hetzel, W.C. (1993) *The Complete Guide to Software Testing*, Second Edition, John Wiley & Sons, USA.

Hirsh, W., King, G., Lovery, J., Fyratt, J. and Hayday, S. (1990) *SUCCESSION PLANNING: Current Practice and Future Issues*. The Institute for Employment Studies, UK.

Holmqvist, M. (2003) Intra- and interorganisational learning process: An empirical comparison. *Scandinavian Journal of Management*, 19, 443–466.

Holt, P., Fontaine, C., Gismondi, J. and Ramsden, D. (1995) *Collaborative learning using guided discovery on the Internet*. <http://ccism.pc.athabascau.ca/html/ccism/deresrce/icce95.htm> [accessed 27/02/2008]

Holt, G.D., Love, P.E.D. and Li, H. (2000) The learning organisation: Toward a paradigm for mutually beneficial strategic construction alliances. *International Journal of Project Management*, 18, 415–421.

Howard, J. (2004) *Demystifying the Extranet, Intranet Journal.* <http://www
.intranetjournal.com/articles/200406/ij_06_08_04a.html> [accessed 04/08/2005]

Huang, J.C. and Newell, S. (2003) Knowledge integration processes and
dynamics within the context of cross-functional projects. *International
Journal of Project Management,* 21, 167–176.

Huang, T.C. (2001) Succession management systems and human resource
outcomes.` *International Journal of Manpower,* 22(8), 736–747.

Huber, G.P. (1991) Organisational learning: The contributing processes and
the literatures. *Organisation Science,* The Institute of Management Sciences,
2(1), 88–114.

Hutcheson, M.L. (2003) *Software Testing Fundamentals.* Wiley, Indiana.

Hutchings, K. and Michailova, S. (2004) Facilitating knowledge sharing in
Russian and Chinese subsidiaries: The role of personal networks and
group membership. *Journal of Knowledge Management,* 8(2), 84–94.

Hwang, A. (2003) Training strategies in the management of knowledge.
Journal of Knowledge Management, 7(3), 92–104.

ILOI – Internationales Institut fur Lernande Organisation und Innovation
(1997) *Knowledge Management – Ein empirisch gestutzter Leitfaden zum
Management des Produktionsfaktors Wissen,* ILOI, Munich.

ITCBP (2003), *Knowledge management.* <http://www.itcbp.org.uk/knowl-
edgemanagement/> [accessed 10/10/03]

Jackson, C. (1998) Process to product: Creating tools for knowledge management, *Proceedings of the 2nd International Conference on Technology Policy
and Innovation, Assessment, Commercialisation and application of Science and
Technology and the Management of Knowledge,* 3–5 August, Lisbon, Portugal.

Jashapara, A. (2004) *Knowledge Management: An Integrated Approach.* Financial
Times Prentice Hall, UK.

Jones, M. (1995) Organisational learning: Collective mind or cognitivist meta-
phor? *Accounting, Management and Information Technology,* 5(1), 61–77.

Kagioglou, M., Cooper, R., Aouad, G., Hinks, J., Sexton, M. and Sheath,
D.M. (1998) *A Generic Guide to the Design and Construction Process Protocol.*
University of Salford, UK.

Kamara, J.M., Anumba, C.J. and Carrillo, P.M. (2002a) A CLEVER approach to
selecting a knowledge management strategy. *International Journal of Project
Management,* 20(3), 205–211.

Kamara, J.M., Augenbroe, G., Anumba, C.J. and Carrillo, P.M. (2002b)
Knowledge management in the architecture, engineering and construction
industry. *Construction Innovation,* 2(1), 53–67.

Kamara, J.M., Anumba, C.J., Carrillo, P.M. and Bouchlaghem, N.M. (2003)
Conceptual framework for live capture of project knowledge, In: R. Amor,
ed.,*Proceedings of CIBW078 International Conference on Information Technology
for Construction,* Waiheke Island, New Zealand, 23–35 April, pp. 178–185.

Kamara, J.M., Anumba, C.J. and Carrillo, P.M. (2005) Cross-project knowledge
management. In: C.J. Anumba, C. Egbu and P. Carrillo (ed.), *Knowledge
Management in Construction,* First Edition, pp. 103–120. Blackwell Publishing
Ltd, Oxford, UK.

Kanter, J. (1999) Knowledge management practically speaking. *Information
Systems Management,* Fall, 7–15.

Kaplan, S. (2002) *Building communities – Strategies for collaborative learning*, ASTD Source for E-Learning. <http://www.astd.org>

Karanikas, H. and Theodoulidis B. (2002) Knowledge discovery in text and text mining software, *Technical Report, Centre for Research in Information Management (CRIM)*, Department of Computation, UMIST, UK.

Kartam, N.A. (1996) Making effective use of construction lessons learned in project life cycle *Journal of Construction Engineering and Management*, March, 14–21.

Kasvi, J.J.J., Vartiainen, M. and Hailikari, M. (2003) Managing knowledge and knowledge competences in project and project organisations. *International Journal of Project Management*, 21, 571–582.

Kerth, N. (2000) The ritual of retrospectives: How to maximise group learning by understanding past projects. *Software Testing & Quality Engineering*, September/October, 53–57.

Khalfan, M.A., Bouchlaghem, N.M., Anumba, C.J. and Carrillo, P.M. (2002) A framework for managing sustainability knowledge: The C-SanD approach, *Proceedings of the European Conference on Information and Communication Technology Advances and Innovation in the Knowledge Society (e-SM@RT) 2002*, University of Salford, Salford, UK, pp. 112–122.

Kleiner, A. and Roth, G. (1997) How to make experience your company's best teacher. *Harvard Business Review*, 75(5), pp. 172–177.

Kleinman, G., Siegel, P.H. and Eckstein, C. (2001) Mentoring and learning: The case of CPA firms. *Leadership & Organization Development Journal*, 22(1), 22–33.

KLICON (2001) *Knowledge learning in CONstruction* (KLICON Project Website at <http://www2.umist.ac.uk/construction/research/management/klicon/index.html> [accessed 07/10/2001]

Kolodner, J. (1993) *Case-based Reasoning*. Morgan Kaufmann, San Mateo, CA.

Koskinen, K.U., Pihlanto, P. and Vanharanta, H. (2003) Tacit knowledge acquisition and sharing in a project work context. *International Journal of Project Management*, 21, 281–290.

KPMG (1998) *Knowledge Management Research Report 1998*, KPMG Management Consulting.

KPMG (2003) *Insights from KPMG's European Knowledge Management Survey 2002/2003*, KPMG Management Consulting, The Netherlands.

Kram, K.E. (1988) *Mentoring at Work: Developmental Relationships in Organizational Life*. University Press of America, Lanham, Maryland, USA.

Kransdorff, A. (1996) Succession planning in a fast-changing world. *Management Decision*, 34(2), 30–34.

Kululanga, G.K. and McCaffer, R. (2001) Measuring knowledge management for construction organizations. *Engineering, Construction and Management*, 8(5/6), 346–354.

Ladd, D.A. and Heminger, A.R. (2002) An investigation of organizational culture factors that may influence knowledge transfer, *Proceedings of the 36th Hawaii International Conference on Systems Sciences (HICSS'03)*, Big Island, Hawaii, 6–9 January. <csdl.computer.org/comp/proceedings/ hicss/2003/1874/04/187440120a.pdf> [accessed 19/12/2003]

Lang, J.C. (2001) Managerial concerns in knowledge management. *Journal of Knowledge Management*, 5(1), 43–57.

Latham, M. (1994) *Constructing the Team – Final Report on Joint Review of Procurement and Contractual Arrangement in the UK Construction Industry,* HMSO, London, UK.

Laudon, K.C. and Laudon, P.L. (2000) *Management Information Systems,* Sixth Edition, Prentice-Hall, New Jersey.

Lazarus, D. and Clifton, R. (2001) *Managing Project Change: A Best Practice Guide,* CIRIA, UK.

Leavitt, P.M. (2003) *The role of knowledge management in new drug development,* American Productivity Quality Centre. <http://www.apqc.org/portal/apqc/ksn?paf_gear_id=contentgearhome&paf_dm=full&pageselect=include&docid=109994> [accessed 19/12/2003]

Leibman, M., Bruer, R.A. and Maki, B.R. (1996) Succession management: The next generation of succession planning. *Human Resource Planning,* 19(3), 16–29.

Leong, E.K.F., Ewing, M.T. and Pitt, L.F. (2004) Analysing competitor's online persuasive themes with text mining. *Marketing Intelligence and Planning,* 22(2), 187–200.

Lessem, R. (1990) *Developmental Management: Principles of Holistic Business.* Basil Blackwell, Oxford, UK.

Levina, N. (1999) *Knowledge and organizations literature review, a report prepared for The Society for Organizational Learning.* <http://pages.stern.nyu.edu/~nlevina/Papers/Knowledge%20Management%20Report.pdf>, [accessed 19/09/2003]

Lewin, K. (1946) Action research and minority problems. *Journal of Social Issues,* 2(4), 34–46.

Lima, C., Zarli, A. and Bourdeau, M. (2002) The e-CKMI: The e-COGNOS infrastructure to support KM in the construction industry. In: Z. Turk and R. Scherer (eds), *eWork and eBusiness in Architecture, Engineering and Construction,* pp. 655–661.

Linton, M. (1975) Memory for real-world events. In: D.A. Norman and D.E. Rumelhart (eds), *Explorations in Cognition,* W.H.Freeman, San Francisco.

Love, P.E.D., Edum-Fotwe, F. and Irani, Z. (2003) Management of knowledge in project environment. *International Journal of Project Management,* 21(3), 155–156.

Mach, M.A. and Owoc, M.L. (2001) Validation as the integral part of a knowledge management process, *Proceeding of Informing Science Conference,* Krakow, Poland, 19–22 June. <http://ecommerce.lebow.drexel.edu/eli/pdf/MachEBKValid.pdf> [accessed 03/08/2004]

Maier, R. (2002) *Knowledge Management Systems: Information and Communication Technologies for Knowledge Management.* Springer, Berlin.

Majchrzak, A., Cooper, L.P. and Neece, O.E. (2004) Knowledge reuse for innovation. *Management Science,* 50(2), 174–188.

Markus, M.L. (2001) Toward a theory of knowledge reuse: Types of knowledge reuse situations and factors in reuse success, *Journal of Management Information Systems,* 18(1), 57–93.

Marshall, L., (1997) Facilitating knowledge management and knowledge sharing: New opportunities for information professionals. *Online,* 21(5), 92–102.

Marshall, N. and Sapsed, J. (2000) The limits of disembodied knowledge: Challenges of inter-project learning in the production of complex products and systems, *Paper presented at the Conference 'Knowledge Management: Concepts and Controversies'*, University of Warwick, Coventry, 10–11 February.

Mason, D. and Pauleen, D.J. (2003) Perceptions of knowledge management: A qualitative analysis. *Journal of Knowledge Management*, 7(4), 34–38.

McCarthy, T.J., Kahn, H.J., Elhag, T.M.S., Williams, A.R., Milburn, R. and Patel, M.B. (2000) Knowledge management in the designer/constructor interface. In: R. Fruchter, F. Peña-Mora and W.M.K. Rodis (eds), *Proceedings of the 8th International Conference on Computing in Civil and Building Engineering*, Reston, USA, 14–16 August, pp. 836–843.

McDaniel, J.G. (2002) Improving system quality through software evaluation. *Computers in Biology and Medicine*, 32, 127–140.

McGee, K.G. (2004) *Heads Up: How to Anticipate Business Surprises and Seize Opportunities First*. Harvard Business School Press, MA.

McGregor, J.D. and Sykes, D.A. (2001) *A Practical Guide to Testing Object-Oriented Software*. Addison-Wesley, USA.

McLoughlin, I.P., Alderman, N., Ivory, C.J., Thwaits, A. and Vaughan, R. (2000) Knowledge management in long term engineering projects, *Paper Presented at the Knowledge Management: Controversies and Causes Conference*, University of Warwick, 10–11 February.

Megginson, D. (2000) Current issues in mentoring. *Career Development International*, 5(4/5), 256–260.

Mentzas, G., Apostolou, D., Young, R. and Abecker, A. (2001) Knowledge networking: A holistic solution for leveraging corporate knowledge. *Journal of Knowledge Management*, 5(1), 94–106.

Mertins, K., Heisig, P. and Vorbeck, J. (2001) *Knowledge Management: Best Practices in Europe*. Springer, New York.

Ministry of the Premier & Cabinet (1999) *Passing the torch: Managing succession in the Western Australian public sector*. <http://www.dpc.wa.gov.au/psmd/pubs/wac/managesucc/success.pdf> [accessed 09/03/2004]

Neve, T.O. (2003) Right questions to capture knowledge. *Electronic Journal of Knowledge Management*, 1(1), 47–54.

Newell, S., Scarbrough, H., Swan, J. and Hislop, D. (2000) Intranets and knowledge management: De-centred technologies and the limits of technological discourse. In: C. Prichard, R. Hull, M. Chumber and H. Willmott (eds), *Managing knowledge: Critical investigations of work and learning*, MacMillan, Wales.

Newell, S., Robertson, M., Scarbrough, H. and Swan, J. (2002) *Managing Knowledge Work*. Palgrave, New York.

Nicolas, R. (2004) Knowledge management impacts on decision making process. *Journal of Knowledge Management*, 8(1), 20–31.

Nonaka, I. and Takeuchi, H. (1995) *The Knowledge-Creating Company: How Japanese Companies Create the Dynamic of Innovation*. Oxford University Press, Oxford.

Nonaka, I. and Toyama, R. (2003) The knowledge-creating theory revisited: Knowledge creation as a synthesizing process. In: *Knowledge Management Research and Practice*, Palgrave Macmillan, Vol. 1, pp. 2–10.

O'Dell, C.S., Essaides, N., Ostro, N. and Grayson, C. (1998) *If Only We Knew What We Know: The Transfer of Internal Knowledge and Best Practice*, Free Press.

O'Leary, D.E. (2001) How knowledge reuse informs effective system design and implementation. *IEEE Intelligent Systems*, January–February, 44–49.

OECD (1996) *The Knowledge-Based Economy*, OECD, Paris.

Orange, G., Burke, A. and Cushman, M. (1999) An approach to support reflection and organisation learning within the UK construction industry, *Proceedings of BITWorld'99: Cape Town*, SA, 30 June–2 July. <http://is/lse.ac.uk/b-hive> [accessed 21/10/2003]

Orange, G., Burke, A. and Boam, J. (2000) The facilitation of cross organisational learning and knowledge management to foster partnering within the UK construction industry, *Proceedings of ECIS 2000*, Vienna, Austria, 5–7 July. <http://is.lse.ac.uk/b-hive> [accessed 21/10/2003]

Ould, M.A. and Unwin, C. (1986) *Testing in Software Development*. Cambridge University Press, UK.

Pakes, A. and Schankerman, M. (1979) The rate of obsolescence of knowledge research gestation lags, and the private rate of return to research resources, *Working Paper No. 346*, National Bureau of Economic Research, Cambridge, USA. <http://www.nber.org/papers/w0346> [accessed 12/07/04]

Panitz, T. (1996) *A definition of collaborative vs cooperative learning.* <http://www.lgu.ac.uk/deliberations/collab.learning/panitz2.html>

Park, M. (2002) Dynamic change management for Fast-tracking Construction Projects, *Proceedings of 19th International Symposium on Automation and Robotics in Construction (ISARC)*, National Institute of Standards and Technology, Maryland, USA, 23–25 September, pp. 81–89.

Patel, M.B., McCarthy, T.J., Morris, P.W.G. and Elhag, T.M.S. (2000) The role of IT in capturing and managing knowledge for organisational learning on construction projects, *The Construction Information technology, International Conference*, Kelandic Building Research Institute, Reykjavik, Iceland, June 28–30.

PEER (2002) *The Bureau of Building's Management of Construction Change Orders – Report #429*, PEER, USA.

Pentland, B.T. (1995) Information systems and organisational learning: The social epistemology of organisational knowledge systems. *Accounting, Management and Information Technology*, 5(1), 1–21.

Polanyi, M.E. (1958) *Personal Knowledge: Towards a Post Critical Philosophy.* Routledge and Kegan Paul, London.

Quintas, P., Lefrere, P., Jones, G. (1997) Knowledge management: A strategic agenda. *Journal of Long Range Planning*, 30(3), 385–991.

Ramaprasad, A. and Prakash, A.N. (2003) Emergent project management: How foreign managers can leverage local knowledge. *International Journal of Project Management*, 21, 199–205.

Reiner, K. and Fruchter, R. (2000) Project memory capture in globally distributed facility design. In: R. Fruchter, F. Peña-Mora and W. M. K. Rodis (eds), *Proceedings of the 8th International Conference on Computing in Civil and Building Engineering*, Stanford University, CA, 14–16 August, pp. 820–827.

Rennie, M. (1999) Accounting for knowledge assets: Do we need a new financial statement? *International Journal of Technology Management*, 18(5/6/7/8), 648–659.

Rezgui, Y. (2001) Review of information and the state of the art of knowledge management practices in the construction industry. *The Knowledge Engineering Review*, 16(3), 241–254.

Ribes, B., Ziemilski, A., Gutelman, M. *et al.* (1981) *Domination or sharing: Endogenous development and the transfer of knowledge*, The UNESCO Press, France.

Rich, E. and Duchessi, P. (2001) Models for understanding the dynamics of organizational knowledge in consulting firms, *Proceedings of the Hawai'i International Conference on System Sciences*, Maui, Hawaii, 3–6 January. <http://www.hicss.hawaii.edu/HICSS_34/PDFs/DTABS05.pdf> [accessed 21/10/2003]

Robertson, M. (1999) Expert consulting: A case of managed autonomy. In: H. Scarbrough, and J. Swan (eds.), *Issues in People Management: Case Studies in Knowledge Management*, Institute of Personnel and Development, London.

Robertson, M., Sorensen, C., and Swan, J. (2001) Survival of the leanest: Intensive knowledge work and groupware adaptation. *Information Technology and People*, 14(4), 334–352.

Robinson, H.S., Carrillo, P.M., Anumba, C.J. and Al-Ghassani, A.M. (2001) Knowledge management: Towards an integrated strategy for construction project organisations, *Proceedings of the 4th European Project Management Conference (PMI)*, Café Royal, London, 6–7 June [published on CD].

Robinson, H.S., Carrillo, P.M., Anumba, C.J. and Al-Ghassani, A.M. (2002) Knowledge management for continuous improvement in project organisations, *Proceedings of 10th International Symposium on Construction Innovation in Project Organisations*, CIB 65, September, Vol. 1, pp. 680–697.

Robinson, H.S., Carrillo, P.M. and Anumba, C.J. (2004) *A Survey of Construction and Client Organisations involved in PFI Projects: A Knowledge Transfer Approach to Continuous Improvement on PFI Projects Report*, Department of Civil and Building Engineering, Loughborough University.

Rollett, H. (2003) *Knowledge Management: Processes and Technologies.* Kluwer Academic Publishers, London.

Roper, M. (1994) *International Software Quality Assurance Series – Software Testing.* McGraw-Hill, Berkshire, England.

Ruggles, R. (1997a) *Knowledge Management Tools.* Butterworth-Heinemann.

Ruggles, R. (1997b) Knowledge tools: Using technology to manage knowledge better, *Working paper for Ernst and Young.* <http://www.businessinnovation.ey.com/mko/html/toolsrr.html> [accessed 02/09/2003]

Ruhleder, K. and Twidale, M. (2000) *Reflective collaborative learning on the Web: Drawing on the Master Class.* Peer Reviewed Journal on the Internet. First Monday, 5(5). <http://www.firstmonday.dk/issues/issues5_5/ruhleder/>

Ruikar, K., Al-Ghassani, A.M., Anumba, C.J. *et al.* (2003) *Techniques & technologies for KM-WP 3 Interim Report.* <http://www.knowledgemanagement.uk.net/pdf-files/KM%20for%20Sustainable%20Const.%20Competitiveness%20WP3%20Interim%20Report-6thmarch.pdf> [accessed 13/10/03]

Ruikar, K., Anumba, C.J. and Carrillo, P.M. (2005) End-user perspectives on use of project extranets in construction organisations. *Engineering, Construction and Architectural Management*, 12(3), 222–235.

Sadler-smith, E., Gardiner, P., Badger, B., Chaston, I. and Stubberfield, J. (2000) Using collaborative learning to develop small firms. *Human Resource Development International*, 3(3), 285–306.

Saint-Onge, H. and Wallace, D. (2003) *Leveraging Communities of Practice for Strategic Advantage*. Butterworth-Heinemann, Oxford.

Schindler, M. and Eppler, M.J. (2003) Harvesting project knowledge: A review of project learning methods and success factors. *International Journal of Project Management*, 21, 219–228.

Schultze, U. (1998) Investigating the contradictions in knowledge management, *Proceeding of IFIP WG8.2 & WG8.6 Joint Working Conference on Information Systems: Current Issues and Future Changes*, Helsinki, Finland. <http://is.lse.ac.uk/helsinki/schultze.pdf>[accessed 21/10/2003]

Shapiro, G. (1999) Inter-project knowledge capture and transfer: An overview of definitions, tools and practices, *Working Paper, CoPS Publication No. 62*.

Shiu, S.C.K. and Pal, S.K. (2004) Case-based reasoning: Concepts, features and soft computing. *Applied Intelligence*, 21, 233–238.

Siemieniuch, C.E. and Sinclair, M.A. (1999) Organisational aspects of knowledge lifecycle management in manufacturing. *International Journal Human-Computer Studies*, 51, 517–547.

Skyrme, D.J. (1998) Knowledge management solutions – The IT contribution, *ACM SIGGROUP Bulletin*, Special Issue on Knowledge Management at Work. <www.skyrme.com/pubs/acm0398.doc> [accessed 06/08/2004]

Smolnik, S. and Erdmann, I. (2003) Visual navigation of distributed knowledge structures in groupware-based organizational memories. *Business Process Management*, 9(3), 261–280.

Soliman, F. and Spooner, K. (2000) Strategies for implementing knowledge management: Role of human resources management. *Journal of Knowledge Management*, 4(4), 337–345.

Sommerville, I. (2001) *Software Engineering*. Addison-Wesley, Harlow.

Spender, J.C. (2008) Organisational learning and knowledge management: Whence and whither? *Management Learning*, 39(2), 159–176.

Stewart, T.A. (1997) *Intellectual Capital: The New Wealth of Organisations*. Doubleday, New York.

Strati, A. (2007) Sensible knowledge and practice-based learning. *Management Learning*, 38(1), 61–77, Sage.

Szulanski, G. (2000) The process of knowledge transfer: A diachronic analysis of stickiness. *Organizational Behavior and Human Decision Processes*, 82(1), 9–27.

Tabbron, A., Macaulay, S. and Cook, S. (1997) Making mentoring work. *Training for Quality*, 5(1), 6–9.

Tan, A. (1999) Text mining: The state of the art and the challenges, *Proceedings of the Pacific Asia Conference on Knowledge Discovery and Data Mining PAKDD'99 workshop on Knowledge Discovery from Advanced Databases*, Beijing, China, pp. 71–76.

Tan, H.C. (2002) The benefits of knowledge management in PFI projects. MSc Dissertation, September 2002, Department of Civil and Building Engineering, Loughborough University, UK.

Tan, H.C., Udeaja, C.E., Carrillo, P.M., Kamara, J.M., Anumba, C.J. and Bouchlaghem, N.M. (2004) Knowledge capture and reuse in construction projects: Concepts, practices and tools, Loughborough University, UK. ISBN: 1 897911 29 7.

Thomas, J.B., Sussman, S.W. and Henderson, J.C. (2001) Understanding 'Strategic Learning': Linking organisational learning, knowledge management, and sensemaking. *Organisational Science*, 12(3), 331–345.

Thomas, J.W. (2000) A review of research on project-based learning. <http://www.autodesk.com/foundation>

Tiwana, A. (2000) The *Knowledge Management Toolkit*. Prentice Hall, New Jersey, USA.

Tiwana, A. (2002) *The Knowledge Management Toolkit: Practical Techniques for Building a Knowledge Management System*. Prentice Hall, London.

Trauner, T.J. (1993) *Managing the Construction Project – A Practical Guide for the Project Manager*. John Wiley and Sons, UK.

Tsui, E. (2002) Technologies for personal and peer-to-peer (P2P) knowledge management, *CSC Leading Edge Forum (LEF) Technology Grant Report*. <http://www.csc.com/aboutus/lef/mds67_off/uploads/P2P_KM.pdf> [accessed 20/10/03]

Tyndale, P. (2002) A taxonomy of knowledge management software tools: Origins and applications. *Evaluation and Program Planning*, 25(2), 183–190.

Udeaja, C.E., Kamara, J.M., Anumba, C.J., Carrillo, P.M., Bouchlaghem, N.M. and Tan, H.C. (2005) Towards a collaborative learning approach for live capture and reuse of knowledge in construction projects. In: S. Sariyildiz and B. Tuncer (eds), *Proceedings of 3rd International Conference on Innovation in Architecture, Engineering and Construction*, Rotterdam, The Netherlands, 15–17 June, Vol. 1, pp. 165–171.

Udeaja, C.E., Kamara, J.M., Anumba, C.J., Carrillo, P.M., Bouchlaghem, N.M. and Tan, H.C. (2008) A web-based prototype for live capture and reuse of construction project knowledge. *Automation in Construction*, Elsevier, 17(7), 839–851.

Venters, W. (2002) *Literature for C-Sand: Knowledge management*. <http://www.c-sand.org.uk/Documents/WP1001-02-KMLitRev.pdf> [accessed 21/10/2003]

Vlahavas, I., Stamelos, I., Refanidis, I., and Tsouki`as, A. (1999) ESSE: An expert system for software evaluation. *KnowledgeBased Systems*, 12, 183–197.

Von Krogh, G., Ichijo, K., Nonaka, I. (2000) *Enabling Knowledge Creation: How to Unlock the Mystery of Tacit Knowledge and Release the Power of Innovation*. Oxford University Press, New York.

Vorbeck, J. and Finke, I. (2001) Motivation and competence for knowledge management. In: K. Mertins, P. Heisig and J. Vorbeck (eds), *Knowledge Management: Best Practices in Europe*, Springer-Verlag, Berlin.

Walsham, G. (2001) Knowledge management: The benefits and limitations of computer systems. *European Management Journal*, 19(6), 599–608.

Watson, I. (1999) Internet, intranet, extranet: Managing the information bazaar. *Aslib Proceedings*, 51(4), 109–114.

Wenger, E. (1998) *Communities of Practice: Learning Meaning, and Identity*. Cambridge University Press, UK.

Wenger, E., McDermott, R. and Snyder, W.M. (2002) *A Guide To Managing Knowledge: Cultivating Communities of Practice*. Harvard Business School Press, Boston, Massachusetts.

Wensley, A.K.P. and Verwijk-O'Sullivan (2000) *Tools for Knowledge Management*, Jospeh L. Rotman School of Management, University of Toronto, Ontario. <http://www.icasit.org/km/resources/toolsforkm.doc> [accessed 20/10/03]

Whetherill, M., Rezgui, Y., Lima, C. and Zarli, A. (2002) Knowledge management for the construction industry: The e-COGNOS Project, *ITcon*, Vol. 7, Special Issue: ICT for Knowledge Management in Construction, pp. 183–196.

Wikipedia (2006a) PHP. <http://en.wikipedia.org/wiki/PHP> [accessed 07/04/2006]

Wikipedia (2006b) MySQL. <http://en.wikipedia.org/wiki/Mysql> [accessed 07/04/2006]

Williams, T., Eden, C., Ackerman, F. and Howick, S. (2001) Management science: Theory, method and practice – The use of project post-mortems, Strathclyde Business School. <ftp://www.managementscience.org/man-sci/papers/wp0107.pdf> [accessed 04/10/2003]

Zahm, S. (2000) No question about it – e-learning is here to say: A quick history of e-learning evolution. *E-learning*, 1(1), 44–47.

Index

Printed and bound by CPI Group (UK) Ltd, Croydon, CR0 4YY

16/04/2025

14658830-0001